Corporeality, Medical Technologies and Contemporary Culture

This book examines the confusions and contradictions that manifest in prevalent attitudes towards the body, as well as in related bodily practices.

The body is simultaneously our reference for the certainties of nature and the locus of transformation and reinvention. The body is at the same time worshipped and despised; an object of desire and of design. Francisco Ortega analyses how the body has become both a screen for the projection of our ideas and imaginings about ourselves and conversely an object of suspicion, anxiety and discomfort. Addressing practices of corporeal ascesis (such as body-building and dietetics), medical technologies and radical anatomical modifications, Ortega documents the ambiguous legacy of a Western theoretical tradition that has always despised the body.

Utilising a theoretical framework that is mainly informed by the phenomenology of the body, feminist theory, disability studies and the thought of Michel Foucault, *Corporeality, Medical Technologies and Contemporary Culture* addresses several ethical and psychological issues associated with the experience and perception of the body in our cultural landscape. Drawing on these diverse areas of philosophical and analytical work, this book will interest those researching law, medicine and sociology.

Francisco Ortega is Associate Professor at the Institute for Social Medicine of the State University of Rio de Janeiro, and Visiting Senior Research Fellow at the Department of Social Science, Health and Medicine of King's College, London (2012–2013).

Corporeality, Medical Technologies and Contemporary Culture

Francisco Ortega

Birkbeck
LAW PRESS

First published 2014
by Birkbeck Law Press
2 Park Square, Milton Park, Abingdon, Oxon OX14 4RN

and by Routledge
711 Third Avenue, New York, NY 10017

A Birkbeck Law Press title

Birkbeck Law Press title is an imprint of the Taylor & Francis Group, an informa business

British Library Cataloguing in Publication Data
A catalogue record for this book is available from the British Library

Library of Congress Cataloging in Publication Data
Ortega, Francisco, 1967–
Corporeality, medical technologies and contemporary culture / Francisco Ortega.
pages cm. — (Birkbeck law press)
ISBN 978-0-415-59322-9 (hardback) — ISBN 978-0-203-79686-3 (ebk) 1. Body image. 2. Human body (Philosophy) 3. Human body—Social aspects. I. Title.
BF697.5.B63O77 2013
306.4'613—dc23 2013022366

ISBN: 978-0-415-59322-9 (hbk)
ISBN: 978-0-203-79686-3 (ebk)

Typeset in Garamond
by Keystroke, Station Road, Codsall, Wolverhampton

For my daughter, Luisa

Contents

List of figures

Preface

'Our body itself is the palmary instance of the ambiguous'

(William James)

William James's remark accurately encapsulates the subject of *Corporeality, Medical Technologies and Contemporary Culture*. This book examines certain confusions and contradictions that manifest themselves in prevalent attitudes towards the body, as well as in related bodily practices. It reflects on such situations and signals different instances of corporeal ambiguity in medical, cultural and legal debates. Why is the body ambiguous? Because it has no intrinsic meaning and because the physiological processes that govern it are subliminal and uncontrollable. But also because the body revealed in the tremendous cascade of scholarship around it, is one that is largely disincarnated, dematerialised and eviscerated. A similar process can be applied to bodies before the law: accounts of corporeality in civil and property law treat the body as divorced from the subject and market alienable and additionally have the capacity to devaluate embodiment and the subject's physical experiences.

This book addresses these corporeal ambiguities and the ways in which they are implicated in practices that affect the body's outer surface, such as diets, fitness and body modifications, and others that concern its invisible interior, such as medical visualisation. These practices and technologies of bodily intervention display the body as a screen on which we project our ego ideals, as our visiting card within the web of social relationships, as the space in which individuals feel able to reinvent themselves. However, they also reveal a certain contempt for corporeality, a desire to transcend materiality and to reject those abject features – blood, viscera, imperfections – that contribute fundamentally to making our bodies real and human. Turned into an object of anxiety, insecurity and discomfort, the carnal body has become a problem both in epistemological debates and in legal encounters. It also rules the lives of many individuals who do not conform to the ideals of perfection and beauty promoted by the diet, fitness and cosmetic surgery industries, and leads to diverse manifestations of body dissatisfaction and body hatred.

This is largely a philosophical book. Its theoretical framework is mainly informed by the phenomenology of the body, feminist theory, disability studies and the thought of Michel Foucault. These perspectives problematise claims that may seem conflicting or paradoxical, and address several ethical and psychological issues associated with the experience and perception of the body in our cultural landscape.

In phenomenological terms, the inside of the body evades a subjective first person perspective. Viscerality constitutes a recessive and absent corporeal dimension – one that eludes direct perception, yet is constitutive of corporeality. Technologies of medical visualisation try to deal with the fundamental ontological ambivalence of the visceral, by transforming the uncanny into the familiar, since viscerality is a challenge to the biomedical project of constituting an objective and causal knowledge of the body. However, such objective knowledge does not coincide phenomenologically with subjective experience. Hence the conflict between the non-objectifiable dimension of viscerality and the biomedical objectification of the lived body, together with the disincarnation of subjectivity that results from those technologies.

From a phenomenological perspective, the reduction of the experience of the lived body to the knowledge of an objective, measurable, quantified and fragmented body, which since the Vesalian revolution accompanies anatomical practices and visualisation technologies, corresponds to the individual's relation to his or her body as something he or she owns, rather than that which he or she *is*. In modernity, to *have* a body becomes more important than to *be* a body; the body-thing or physical body overshadows the lived body *I am*. The Western anatomical tradition and the history of body visualisation attest to a significant weakening of the latter, associated with the body-machine, the body-corpse or the virtual body.

Similarly, the corporeal metaphors underlying legal narratives – the body as machine, property, commodity or object of privacy rights – identify our relationship to our bodies with our connection to material possessions; they thereby contribute to embedding the objectification and commodification of bodies further within Western philosophical thinking, the history of medical visualisation, and the consumer culture of healthism and bioascesis that will be examined in this book. Property relationships, as presupposed in the legal notion of 'bodily integrity', lead to the disembodiment of the legal subject, the indifference towards individuals' bodily nature and corporeal experiences, and the total commodification of human personhood. And yet, human bodies are not monadic, isolated and independent entities. They exist in a state of deep dependency and interconnectedness with other bodies and the surrounding environment throughout human life; this fact runs against notions of corporeality as a self-sufficient entity. Prosthetics, organ transplantation and other forms of body modification defy ideas of a 'natural' body, as well as the normative structure of human embodiment assumed both in property law and the history of medical visualisation.

These are all insights also present in the phenomenology of the body and other philosophical and psychological theories discussed in this book, which may be useful tools to deconstruct the legal status of the body based on property, right to privacy or integrity. Theoretical models which foreground the corporeal nature of the legal subject dislocate the property paradigm and its dualist underpinnings, and take into account the differences in bodily make-up of the subjects before the law. In this book, phenomenology, disability studies and critical legal studies inform such a theory of corporeal justice, and throw critical light on notions of efficiency, productivity, autonomy, independence, rational agency, free will and individualism underlying neoliberal understandings of justice and citizenship. They place the concrete, fleshly and lived existence of the individuals and their embodied experiences at the core of legal narratives and practices and thereby destabilise normalising ideas of corporeality and ableism.

This book critically discusses certain appropriations of Foucault's terminology of ascesis and technologies of the self to examine neoliberal regimes of self-care. Ancient ascesis had a fundamental socio-political dimension; its ultimate objects were the city and other individuals. I will argue that the identification of ancient ascesis with neoliberal regimes of self-care depletes the former of its ethical significance and downplays the ethical and political implications of contemporary practices of the self. To highlight the difference, I will advance the notion of bioascesis, understood as the wide array of discourses and practices individuals in neoliberal societies employ to think, speak and act upon themselves in the pursuit of the perfect health and body. Bioascesis stresses corporeal, hygienic, medical discourses and practices in ways that are consistent with the neoliberal ideology of the free and autonomous entrepreneurial self. 'Healthism' will be considered as a privileged bioascetic site, where hedonistic lifestyles and the pursuit of the perfect body merge. The narcissist and individualist character of such practices provides an ideal contrast with ancient ascesis.

Against the sociopolitical background of the 'intimate public sphere' (Berlant) encompassing innumerable private lives on display for mass consumption, it may be argued that the private bodies of healthism and bioascesis have become public. The transformation of the public sphere into the intimate public sphere of healthism give visibility to private and apolitical bodies – apolitical in the sense that individuals are more concentrated on a self-centred pursuit than on the betterment of society. The result is an 'anti-political politics' in which diversity is transformed into a conglomerate of apolitical bioidentities.

I do not, however, wish to establish a harsh and artificial dichotomy between ascesis and bioascesis. The book examines practices of body modification that, at the level of an embodied phenomenology, may represent a bulwark against the drive towards body normalisation within healthism and bioascesis. The chapter on law and corporeality further examines how health movements and diverse forms of biological and disabled activism collapsed the public-private divide by turning corporeal features into the basis for rights, inclusion, entitlements and protection – even though those features radically depart from the aesthetic cult of the body advanced by healthism and body consumerism.

This broad cultural landscape is analysed in the book in the following way. I will start by making the theoretical framework explicit, and by examining social constructionist claims about corporeality and contrasting them with the phenomenology of the body and its consequences for understanding the body's role in contemporary culture (Chapter 1). I will then use a phenomenological approach to critically address the disincarnation and objectification present in the history of anatomical dissection and in medical visualisation technologies (Chapter 2). The chapter that follows compares ancient ascetic practices and contemporary bioascesis, examines the main elements of the latter and their ethical implications, and proposes an embodied phenomenology of body modification as resistance to corporeal normalisation (Chapter 3). I conclude by analysing several instances in which the bodily existence of individuals is incorporated in the legal process, as well as the contribution of disability studies to the development of a theory of corporeal justice (Chapter 4).

Acknowledgements

Research for this book was at different stages supported by the Brazilian National Council for Scientific and Technological Development (CNPq), the Carlos Chagas Filho Research Support Foundation of the State of Rio de Janeiro (FAPERJ), the German Academic Exchange Service (DAAD) and the Coordination for the Improvement of Higher Level Personnel (CAPES).

I wish to thank my colleagues and students at the Institute for Social Medicine of the State University of Rio de Janeiro, and especially Jurandir Freire Costa, Benilton Bezerra Jr and Rafaela Zorzanelli. Many ideas in this book were discussed with them during the past years. The Max Plank Institute for the History of Science, Berlin, where I have been a Visiting Scholar several times since 2003, was an outstanding place to undertake research. I am also thankful to Lorraine Daston, director at the Institute, for her hospitality, as well as to the Institute's librarians for their helpful assistance.

My friend Fernando Vidal commented in detail on some of the chapters and always provided salutary intellectual challenge. Sally Eales also provided excellent editing.

The final version of this book was written at the Department of Social Science, Health and Medicine of King's College, London, where I was Senior Visiting Research Fellow in 2012–2013. This intellectual and supportive environment was an ideal place to finish the manuscript. I want to thank especially Nikolas Rose for his hospitality and our conversations in the last few years.

I am also grateful to several individuals who in different ways contributed to this book, including Pietro Barbetta, Jean-François Bissonnette, Costas Douzinas, Pepe Duran, Gabriella Erba, Mechthild Fend, Valeria Gennero, Patrick Hanafin, Rafael Huertas, Hanna Kienzler, Erika Manserus, Sam McLean, Fernanda Nunes, Luís Henrique Sacchi dos Santos, Enrico Valtellina and Francisco Vázquez Garcia.

Susan Aldworth kindly allowed me to use her wonderful artwork for the cover, and the Wellcome Library, London, gave permission to use their images.

As always, I am indebted to my wife Bethania for her love and support.

London, April 2013

Chapter 1

The body between constructionism and phenomenology

. . . and when the word 'material' came out of my mouth it was as though I had farted in public

(Susan Bordo, 'Bringing Body to Theory')

Constructionism and phenomenology

This book addresses central medical, cultural and legal debates around corporeality. Before dealing specifically with those debates, I want in this chapter to make explicit the theoretical framework which will be elaborated in the following ones. Although I favour phenomenological approaches to corporeality, my philosophical position is eclectic. Scholars seldom draw on phenomenological accounts when dealing with the socio-cultural, legal and technological dimensions of corporeality. Their perspectives, for example poststructuralism, or gender and postcolonial studies, tend to be informed by what can, in general terms, be characterised as a constructionist position. I do not think that phenomenology should replace them, but it can supplement them in important ways, and help bring to light their chief epistemological and ethical assumptions.

This chapter critically examines one of those assumptions, namely that constructionist approaches to the body have an intrinsically emancipatory potential. This conviction is shared by movements that have in recent decades placed the body at the core of socio-political debates – movements associated among others with women, homosexual people, persons with disabilities and racial minorities. Realising that basic beliefs about the body are social and political constructs has contributed to undermine oppressive claims about gender, disability, age or ethnicity. Constructionism therefore possesses political and ethical significance. However, without downplaying its liberating potential, I will here examine its problematic epistemological and ontological affinities (as far as matters of corporeality are concerned) with the discourse of biotechnologies.

I am not the first to notice these affinities. Genetic engineering, synthetic biology, AI, organ transplantation, stem-cell research, cloning and medical imaging have all been said to undermine notions of an ideal, natural body, and to highlight its social, economic, political and technological construction (Balsamo 1999; Clough 2012; Petersen 2007). Yet, as we will see in the chapter on law and the body, social constructionist explanations may be appropriated by groups such as global warming sceptics (Latour 2004). Moreover, constructionist claims may involuntarily justify the commodification and fragmentation of contemporary bodies. By emphasising the limits of corporeal malleability, phenomenology may offer an alternative.

The relationship between body fragmentation and body totality, which is explored in the next chapter, mirrors the constructionism/phenomenology dichotomy. As we shall see, the fragmentation of the body operated in the anatomical tradition and in medical visualisation technologies converges with the praise of the fragment within postmodern and poststructuralist thinking, which refuses all forms of totality. Similarly, constructionist arguments around the body are mobilised to counter reified, essentialised and naturalised notions frequently identified not only with physicalist explanations of human corporeality but also with the phenomenological approach. Hence, the same kind of epistemological and ethico-political arguments serve to sustain constructionism and fragmentation. Advocates of both believe their positions offer ethical and epistemological advantages over materialist and phenomenological accounts that emphasise the experience of the lived body. On the contrary, I will argue that a position which emphasises the incarnate and material dimension of corporeality does not imply an essentialist or foundationalist attitude, and offers ethical and epistemological advantages over constructionist perspectives that end up implicitly reproducing the dualisms they explicitly reject.

Why the body now?

In the introduction to the *Routledge Handbook of Body Studies*, Bryan Turner (2012a) traces the increasing theoretical interest in the body displayed in the social sciences and humanities of the last three decades. This growth of attention is related to biomedical and biotechnological innovations, as well as to the rise of social movements that gave prominence to the political and social dimensions of corporeality. The *Handbook* provides a good idea of the size of the literature, and other examples could be given. Maren Lorenz's *Corporeal Past: Introduction to the History of the Body*, published in German in 2000, reviewed 1156 relevant books, chapters and articles (Lorenz 2000). And the collective volume *Medicine in the Twentieth Century* (Cooter and Pickstone 2003) contains a whole section 'Bodies', made up of 17 chapters about the historiographical, healthy, industrial, third-world, temporal, sexual, reproductive, psychological, psychoanalytic, psychiatric, diseased, disabled, defended, genetic, analysed, experimental, ethical and dead body.

The socio-cultural and historical interest in corporeality is, as could be expected, correlative to the appearance of the 'somatic culture' explored in other chapters of this book. Addressing the status and construction of corporeality has become commonplace in the human sciences. This can be traced back to the 1980s, with the publication of such works as Bryan Turner's *Body & Society* (1984), Michel Feher's *Fragments for a History of the Human Body* (1989), Emily Martin's *The Woman in the Body* (1987) or Barbara Duden's *The Woman Beneath the Skin* (published in 1987 in German, in 1991 in English), as well as to the later creation of the journal *Body and Society* in 1995 (Petersen 2007; Turner 2012a); in 2002, political scientist John Tambornino (2002) explored the 'corporeal turn' within social theory in the framework provided by the previous two decades of thematisation of body issues.

The debate on the epistemological status of corporeality in the human sciences has gone through different phases. In the early 1990s, discussions were frequently pervaded by constructionist arguments. Thus, Michel Feher considered the history of the body 'not so much the history of its representations as of its modes of construction' (Feher 1989: 11; see also Bynum 1995; Lorenz 2000; Porter 1992; Stoff 1999; Turner 2012a). In feminist theory, debates around corporeality were mainly informed by a linguistic turn. 'Ironically, although there has a been a tremendous outpouring of scholarship on "the body" in the last twenty years', write Stacy Alaimo and Susan Heckman in their introduction to *Material Feminism*, 'nearly all the work in this area has been confined to the analysis of discourses *about* the body' (Alaimo and Heckman 2008: 3). Similarly, in an earlier volume on *Feminist Theory and the Body*, the editors noticed an 'indifference to materiality', which they considered typical of the 'masculine forms' shared by several postmodern and post-structuralist theories (Shildrick and Price 1999: 7). The loss of physicality is considered an excessively high price to pay for the gains of the linguistic turn in research on the body (Hekman 2008: 88).

Consequently, some scholars now call for a corporeal turn within feminism (Howson 2005), for a 'material turn' or 'material feminism', also labelled 'new materialism' – a trend represented by authors including Sandra Harding, Donna Haraway, Elizabeth Wilson, Karen Barad, Susan Squier and Melissa M. Littlefield (Alaimo and Heckmann 2008: 6–7, 17; see also Alaimo 2008: 242–44; Blackman 2008; Diprose 2012). The material turn also includes scholars such as Elizabeth Grosz, Rosi Braidotti, Susan Bordo and Iris Marion Young, some of them represented in *Material Feminism*, who have drawn on psychoanalytic, phenomenological, Deleuzian and, more recently, Darwinian approaches to counter the prevalence of social constructionism within feminist theory (Bordo 1993; Braidotti 2002; Grosz 1994, 2004; Young 2005).

A parallel instance of the material turn in the humanities and social sciences consists of the growing success of cognitive and neuroscientific methodologies. Neurolaw, neuroanthropology or neuroesthetics, among other *neuro*disciplines (Ortega and Vidal 2011, 2013; Vidal 2011), are supposed to revitalise the corresponding humanistic disciplines. To the extent that the corporeal and material dimension of human existence has become a central concern for the human sciences, social theories, as Nikolas Rose has pointed out (2013: 4), are no longer 'thought progressive by virtue of their distance from the biological. Indeed the reverse assumption is common – it seems that "constructivism" is passé, the linguistic turn has reached a dead end and a rhetoric of materiality is almost obligatory'. In a similar vein Ilina Singh explores the contributions of social science for a better comprehension of the nature-nurture debate in human development, and concludes that 'perhaps one day the long walk from the science lab to the sociology department will have become a stroll down the hall' (Singh 2012a: 319).

Obviously, not everyone feels comfortable with such developments. For instance, regarding specifically the sociology of the body and the material turn in feminist philosophy, sociologist Anne Witz (2000: 2) writes that the biological body purportedly recuperated by the newer materialist theories is 'in danger of being simply the *abject* male body, not the *excluded* female body'; this would make women the 'absent presence' within those theories. Pace Rose and material feminists, constructionist approaches still have currency and significant defenders in the social sciences and humanities (Petersen 2007; Weinberg 2012).

Independently of the future of social constructionism, the debates concerning it serve well the purpose of this chapter and spell out the philosophical orientation of this book. Phenomenology is here the favoured, although not the only, theoretical tool to address the medical, cultural and legal problematisations of corporeality. A contrast with social constructionism helps to elucidate the strengths and shortcomings of the phenomenological perspective. Obviously both standpoints represent the ends of a spectrum which is richer and offers many intermediate positions. So my intention is not to polarise the discussion. In the next sections, I will first examine social constructionist claims about corporeality, then the phenomenological perspective and, finally, the epistemological affinities between social constructionism and biotechnologies. The inevitable starting point, however, is Michel Foucault, whose perspective on corporeality has been decisive for social constructionism (Cooter 2010; Petersen 2007; Turner 2012a, 2012b; Weinberg 2012).

Foucault and the constructed body

Foucault's aversion to the triad phenomenology-Marxism-existentialism is well attested. For him, those positions were philosophies of the conscious subject, and illustrated a foundationalist project he rejected. 'Can it be said', he asked, 'that the subject is the only form of existence possible?' (Foucault 1991: 49). Such a question, from a 1978 interview, summarises Foucault's attitude vis-à-vis a subject who, according to phenomenology and existentialism, gives sense to the world. It is relevant to note that Foucault's critiques of phenomenology in interviews were aimed more at Sartre than at Merleau-Ponty. The latter is criticised in *The Order of Things*, which characterises the existential phenomenology of the body as the project that examines 'how what is given in experience and what renders experience possible correspond to one another in an endless oscillation' (Foucault 1994a: 336).

The body here emerges as an ambiguous entity – both a fact and the condition of possibility of all facts. The interest of Merleau-Ponty's philosophy for our topic resides precisely in that ambiguity, which was for the philosopher constitutive of our experience of corporeality as reversible flesh, that is, we feel our body not merely as being composed of sensible qualities, but also as something that perceives; not only as that which is seen or heard, but as that which sees and hears; not just as tangible, but as actively touching – simultaneously subject and object of experience (Cohen 1984; Merleau-Ponty 2002). The phenomenology of the body would thus transcend the dichotomy of subject and object.

Foucault formulates his constructionist approach to the body in the 1970s, in connection with what is known as the genealogy of power, and it is through this notion that it has inspired more or less radical forms of constructionism. In 1971, in his programmatic 'Nietzsche, Genealogy, History', the French philosopher writes: 'Nothing in man – not even his body – is sufficiently stable to serve as the basis for self-recognition or for understanding other men' (1998: 380). Such assertion brings to light the nominalism and radical constructionism of his view of the body at the time. The radical heterogeneity of bodies implies the impossibility of defining a human 'us'. This has significant consequences.

On the one hand, identities based on biological or physical features may benefit ethnic or social minorities, since they may become the basis of claims for equal treatment; at the same time, as racism and sexism suffice to show, the same features frequently lead to exclusion and discrimination. Foucault was always critical of identity politics, and his observation quoted above anticipates positions he defended in the 1980s, especially in relation to the socio-political role of the gay community (Ortega 1997). His caution in this domain is understandable. On the other hand, his nominalism is less so. The representation of 'us' need not lead to the exclusion of the 'others'. And it is difficult to imagine coherent images of 'us' and 'others' that are not rooted in the comparison between bodies. Foucault sought to criticise Lacan's ideas on the mirror stage, initially formulated in the mid-1930s (Lacan 2001), according to which one's body image is based on the comparison with the other's body – a pattern that (in Lacan's later thought) becomes a permanent structure of subjectivity.

Foucault's text of 1971 introduces the notion of corporeality that will be elaborated in *Discipline and Punish* (1995) and other texts of the 1970s. In his book on 'the birth of the prison', the body is described as infinitely pliable in the hands of power. The object of discipline is the creation of 'docile bodies', which are offered as 'natural bodies' to analysis and manipulation. The individual and his soul are produced by disciplinary operations on the body, which is itself socially constructed (1995, 1980: 73–74). What, however, does it mean exactly to claim that the body is socially constructed?

Social constructionism and the body

In his classic *The Social Construction of What?* (1999) philosopher Ian Hacking brings to light certain confusions inherent in constructionist theses. When we say that X is a construction, where X is for example homosexuality, race, gender, body or any other category or object, the epistemological and ontological status of X generally remains unclear. In relation to the construction of the body, the reference may be for example to one's own body, to the idea of body or to the medical and juridical institutions that define the notion of a natural gendered body. According to Hacking, the confusion results from the fact that the object and the idea of the object are not properly differentiated. For Bruno Latour (2003), in contrast, Hacking's strict distinction between objects and ideas, facts and propositions, reduces the debate to a choice between realism and nominalism, and thereby excludes mediations between words and things.

We may take as example the notion 'natural body' (Petersen 2007: ch 2), as it emerges in the history of 19th century medicine, especially physiology and anatomy. As we shall see in the following chapter, the natural body had at the time the status of 'epistemic thing' (Rheinberger 1997), that is, a material entity defined by what one does not yet know, a projection rather than a definite object. Characterising the natural body as an epistemic thing requires distinguishing two different epistemological levels. One thing is to say that the body was constructed as an entity within the history of medicine. Science is here envisaged as a situated practice, and the historicity of knowledge excludes making a radical distinction between nature and culture. Another thing is to claim that the body's materiality, as well as its phenomenological, first-person dimension are constructed. According to a thoroughgoing constructionism, 'construction' applies not only to discourses about the body, to ideas, theories or classifications about it, or even to how bodily experiences are signified – but also to the body itself, in its very materiality.

Thus, an important distinction to be made in discussions about the body as constructed is that between epistemological postulates and ontological-anthropological assumptions (Landweer 2002) – in other words, between our access to a certain reality and the mode of existence of that reality itself. According to epistemological constructionism, there is no 'natural' access to the body, no knowledge of the body independently of historically relative discursive and social practices. This is an agnostic position, which refuses to make universal claims about human corporeality, and insists on dealing with it in particular contexts, through historical, sociological or anthropological analysis. Such 'weak' form of constructionism does not conflate ontological and epistemological matters (Williams 2003: 25); as Elizabeth Wilson (2008: 377) observes, it is just another way of referring to the social, cultural and historical contexts in which individuals are embedded (see also Siebers 2001: 738–9).

In contrast, the strong version of contructionism does conflate epistemological and ontological matters. It postulates a negative ontology, according to which the body and materiality do not exist as such, but are social constructions and discursive effects. This position is sometimes identified with Judith Butler (1990, 2011). For the feminist philosopher, the body's materiality does not precede discourse, but is 'power's most productive effect'. Reality is performatively generated; and performativity is understood as the 'reiterative and citational practice by which discourse produces the effects that it names' (Butler 2011: xii). To speak of materiality is to speak of the historical processes that bring about materialisation; the body's physicality, therefore, 'should surely be an *object* of feminist inquiry, but . . . would be quite problematic as a *ground* of feminist theory' (Ibid: 22).

Butler's notion of corporeality has been heavily criticised within feminist thinking. Claire Colebrook (2008: 69) speaks of the 'scandal of Butler's linguisticism' and Karen Barad (2008: 151, note 26) laments that in Butler's theory, matter is ultimately reinscribed as a passive product of discursive practices. Elizabeth Grosz argues that, for Butler, mattering is more important than matter or materiality (Ausch, Doane and Perez 2000). German historian Barbara Duden (1993: 28) is even more radical: 'no deconstructivist will talk me out of my corporeality'. Butler has in turn rejected these readings of her work, and dismissed as 'scandalous' the 'strong version' of constructionism, which, as she writes, claims that 'the body is somehow reducible to the linguistic coordinates by which it is identified and identifiable, as if there is no non-linguistic stuff at issue' (Butler 1997: 3; for less negative assessments of Butler, see Petersen 2007; Weinberg 2012).

In Butler's view, to claim that materiality is an effect of discourse/power does not imply that human corporeality is reducible to words, but simply that any understanding or apprehension of the body is mediated by discursive contexts, and that the body does not by itself 'strictly dictate what can be truthfully said about it' (Weinberg 2012: 152; see also Shildrick and Price 1999: 7).

Most constructionists adhere to the agnostic position. However, insofar as such position functions as a metatheory and a methodological postulate (Landweer 2002), it operates like radical constructionism's negative ontology. According to Susan Bordo (1998: 89), all constructionisms are 'discourse foundationalist'. Insights into our 'embeddedness in discourse', she writes, function for them 'as a "bottom line", a privileged framework which is used to deconstruct other frameworks of understanding to its own preferred elements'. That is why she speaks of constructionism as a 'theoretical pasta-machine' that transforms everything into a 'trope' and, having done this, rejects it 'as so much detritus'.

What would Foucault's position have been in this debate? He might have been agnostic, although sometimes oscillating towards negative ontology, as when he asserted that discipline produces the natural body. Several times, he asserted that his position was more materialist than Marxism, since it examined the effects of power not through ideology, but directly in bodies. The individual was for him the 'product of a relation of power exercised over bodies', and power relations, he wrote, 'can materially penetrate the body in depth' (Foucault 1980: 74, 186). In spite of such apparent emphasis on materiality and on the reality of power's effects, Foucault's commitment to the body coexisted, paradoxically, with a position in which it disappears as material or biological entity, and remains knowable only through the filter of power relations as embodied in medical, juridical or pedagogical discourses. The biological body is a manifestation of the social, and plays no role as an active and intentional agent.

Hence Foucault's failure to take into account the body's materiality (Barad 2008; Hughes 2008; Lorenz 2000; McNay 1991; Outram 1989; Shilling 2012; Turner 2001; Turner 2008). As Bryan Turner observed, Foucault simply 'ignores the phenomenology of embodiment' (2008: 206). Except for pleasure and desire, the sensory experiences involved in the fact of having a body plays no role in his analyses, and vanishes behind the norms and practices that regulate it. Power relations transform bodies in objects of knowledge, in symbols, metaphors and sites for the exercise of power, at the expense of bodies as sources of experience. The body thus appears as a metaphysical entity, a surface on which social discourses and practices inscribe themselves via the school, the hospital, the prison, the military barracks and the psychoanalyst's office.

But what does that body feel? Foucault has nothing to say about that, nothing about the feelings of the subject who is disciplined or under the panopticon's gaze. Disciplined bodies seem to feel no fear, frustration, pain, shame, humiliation, unhappiness or anxiety. As Terence Turner (2001: 36) put it, 'Foucault's body has no flesh'. For the French philosopher, as for other constructionists, the body is above all a symbolic construct whose history is consubstantial to that of its representations. When the subjective experience of the body is thus minimised, the body seems to be an indefinitely plastic and available entity. This has ethical and epistemological consequences that will be examined below.

Foucault's disembodied view of corporeality goes hand in hand with the pre-eminence he gives to vision and visual 'apparatuses' (*dispositifs*). Both in *The Birth of the Clinic* and in *Discipline and Punish*, 'gaze' is a fundamental concept. 'This book', wrote Foucault (1994b: ix) in opening *The Birth of the Clinic*, 'is about space, about language and about death; it is about the act of seeing, the gaze'. Both the clinical gaze and the panopticon make the body visible. The rupture between modernity and the classical age is interpreted in visual terms. The modern model of clinical anatomy, which reveals the interior of the body, has a genealogical counterpart in the replacement of the physical means of classical punishment, which were inscribed in the condemned person's body, by the new means of penal justice. The modern 'incorporeal penalty' (Andrieu 2004) tends to substitute vision for touch: 'the guillotine takes life almost without touching the body' (Foucault 1995: 13).

As will be discussed in the next chapter, vision disembodies and dematerialises. It distances us from objects and disconnects sensation from movement (Jonas 2001; Leder 1990; Todes 2001). In contrast, touch affords an experience of reality and materiality, creating a link of co-presence with things. Analogous on this point to visualisation technologies, the devices for construing the body visually offer a model of disembodied corporeality. Such an understanding of the body, which is current in several forms of social constructionism, may be compared to what feminist thinkers Julia Kristeva, Elizabeth Grosz and Gail Weiss, drawing inspiration from Lacan, conceptualised as 'abjection' (Grosz 1990; Kristeva 1982; Weiss 1999a, 1999b; see also Covino 2004). Abjection is here understood as the 'corporeal refusal of corporeality' (Weiss 1999a: 42) or, in Grosz's words, as 'the paradoxically necessary but impossible desire to transcend corporeality' (quoted in ibid: 41).

The abjection of materiality characteristic of the radical versions of social constructionism has an empirical counterpart in the desire to overcome organic physicality that inspires various projects in virtual reality, artificial intelligence and transhumanism (Wolfe 2010; Halberstam and Livingston 1995; Toffoletti 2007). Abjection refers to things related to animality, death and viscerality, which escape symbolisation and threaten ideal body images: flesh, blood, secretions, faeces, urine, sweat, saliva, vomit and other bodily fluids. These are all parts of myself which must be expelled so that I can inscribe myself in the symbolic order. Such mode of inscription, which transforms the body in the discursive entity of social constructionism, requires abjection, the impossible carnal rejection of corporeality.

The phenomenological body

The phenomenological body represents the opposite of the abject body. It does not, however, coincide with the anatomo-physiological body; nor is the material dimension of corporeality reducible to anatomical structures and physiological mechanisms. In saying that phenomenology does not imply physical reductionism, I am indebted to a number of authors who share an emphasis on the role of the environment in the making of the bodily self, on the confluence of the experiences we have of self and world.[1] Obviously, they differ considerably among themselves. Nonetheless, what is significant for us is the fact that they all postulate an intentional link between the body and the physical, social and cultural environment – a link which precludes reducing the former to its anatomo-physiological dimensions. My use of 'phenomenological body' refers to that shared assumption; it is therefore general, and does not adhere to any author in particular.

The phenomenological tradition uses the German terms *Körper* and *Leib*, usually translated as *physical body* and *lived body*, to distinguish the body as construed and described by science from the individual bodily experience.[2] Although the terms are already found in Hegel's *Phenomenology of Mind* (Russon 2001), the difference was conceptualised by Edmund Husserl (1952) and Maurice Merleau-Ponty (2002). This vocabulary does not imply any kind of dualism, but rather designates different dimensions of corporeality. *Körper* is the main historical referent of medicine and the sciences of the body; it is the body fragmented to be measured and quantified. Since the Vesalian revolution of the Renaissance, as we will see in the following chapter, it has remained a constant of anatomical practices and medical visualisation technologies, giving rise to the fiction of a natural body and to the pre-eminence of *Körper* over *Leib*. Both dimensions are constitutive, and the obfuscation of the physical body in some phenomenological writings derives from the fact that they use the lived body to counter the conceptual and social instrumentalisation of the physical body (Schürmann 2003).

Incorporating the phenomenological body into Foucault's approach would, first, do away with the dualism of his thinking; secondly, enable a notion of action that overcomes the impasses of his genealogy of power and allow for resistance; and, finally, explain how power relations operate. This applies not only to Foucault's thinking, which I will here take as paradigmatic object, but also to the various forms of social constructionism.

Transcending dualism

By privileging the body in the formation of the self, Foucault inverts the traditional dualism of soul and body or soul and mind. His thinking nonetheless implies a certain dualism, since he considers the body as separate from the mind.[3] As Axel Honneth (1993: 168) noted, Foucault was less interested in 'historical psychology' than in a 'historical physics' focused on the disciplining of bodies, at the expense of the phenomenological body as a unity integrating physical and psychological processes. In Foucault's description, bodily discipline is not, as in Norbert Elias's *Civilizing Process* (1939) or in Adorno and Horkheimer's *Dialectic of Enlightment* (1947), a psychophysiological process of increased control over the body and the affects. Rather, it is a purely physical process of control of bodily movements. Foucault's aversion to psychology and psychoanalysis led him completely to neglect the individual psyche, and to reduce interiority to that which power and its coercive mechanisms produce inside bodies. Honneth (1993: 27) sees such 'theoretical ambiguities' as vestiges of structuralist notions: the idea of the human body as product of power *dispositifs* is the sociological correlate of the structuralist view of the subject as product of language. Such dualism persists into the last phase of Foucault's *oeuvre*. Thus, speaking of the care of the self in Greek culture, the philosopher refers to the soul as 'subject of all these bodily . . . actions' (Foucault 2005: 55).[4]

A different notion of corporeality would allow us to transcend the Foucauldian and constructionist dualisms. Why would that be important? Mainly, because of its ethical and socio-political consequences. Dualism is an obstacle to examining the effects of normalising power on the body, which was eventually Foucault's aim (Heyes 2007: 23). In what sense does the phenomenological conception of the body overcome dualism?

First, for the various strands of the phenomenology of corporeality, body and mind are not ontologically distinct entities, nor is there (save certain pathologies) a split between the bodily and the mental or a 'place' where body and mind connect. The mental and the physical are manifestations of the same body (Gallhager and Zahavi 2008). The division between the two is a practical way of talking, a tool for carrying out different tasks in different contexts. Subject and object, body and mind, thought and thing, mental and physical are all functional distinctions, made for practical purposes. In other words, we establish a *phenomenological difference* between mental events and physical objects – not an *ontological difference* between two independent metaphysical entities.

Secondly, Foucault's oft-repeated statement: 'Where there is power, there is resistance' (1978: 95), calls for a non-dualist notion of corporeality capable of explaining resistance. Since his emphasis on the effects of power on the body reduces social agents to passive bodies, it makes no room for individuals' autonomous action. Although in *The History of Sexuality: An Introduction* (1978) there is mention of an alternative economy of bodies and pleasures as a form of resistance to biopower, subjects are not given the means to oppose the power that shapes them. In other words, the very same body that is passively constructed by power must become the active source of resistance to that power. Foucault was conscious of that difficulty; the new directions of his thinking starting in the mid-1970s may have been an attempt at leaving the vicious circle inherent in his analysis of power and resistance (Ortega 1997).

For that, however, a different notion of corporeality is necessary. As some commentators have pointed out, there is a contradiction between the body as a product of discourse and power relations on the one hand, and on the other as pre-discursive origin of pleasure and resistance (Dreyfus and Rabinow 1983; Hoy 1999; Turner 2001; Williams and Bendelow 1998). Several feminist scholars criticised the late Foucault: in making room for a pre-discursive libidinal multiplicity and therefore for a sexuality that antecedes the law, he seemed to abandon the constructionist model and embrace a natural body, unmediated by culture and history (Butler 1989; Grosz 1994; McLaren 2002). Endowing the body with the capacity to transcend its discursive and constructed dimension moved Foucault closer to phenomenology.

Indeed, the phenomenological body opens the way to overcoming the idea that corporeal practices are merely instrumental, as happens in Foucauldian 'discipline'. The position of the French philosopher in this respect remained ambiguous, since the kind of corporeality he attributed to the subject of Antiquity who constituted himself as moral subject through bodily practices was not the same as the modern subject's docile and disciplined body. He no longer depicted the passive construction of subjectivity, but an agency involving the notions of freedom and autonomy. This implied a different type of body. The body of Greek and Roman ascesis, which I will address in Chapter 3, does not correspond to the 'somatic singularity' (Foucault 2006a) of disciplines.[5] Since Foucault did not conceptualise this other form of corporeality, we may imagine it is the same as the disciplined body of modernity. But the body that is the source of modern disciplined subjectivities does not fit with the active making of Greco-Roman subjectivities, elements of which may help to articulate resistance in the contemporary context (Ortega 2001). Foucault gives the impression of not wanting to deal with that body, even though it is implicit in the active subjectivity he describes in connection with the Greco-Roman 'aesthetics of existence'. The result is a radicalisation of the dualist position, as seen for example in the notions of *chresis* (use) and *chresis aphrodision* (use of pleasures), which imply that the soul acts on the body and is the true subject of corporeal actions (Foucault 1990, 2005).

Yet there are moments when Foucault seems to get closer to the phenomenological body. Let us mention in the first place a little-known 1966 radio lecture on the 'Utopian body'. It was given in the same year as the publication of *The Order of Things*, in which, as we saw, Foucault criticised the phenomenological notion of corporeality. In the 1966 lecture, however, he made some interesting observations on the phenomenological body. He explicitly drew on the Husserlian notion of the body as 'zero point' of the world (Foucault 2006b: 233):

> . . . it is around it [the body] that things are arranged. It is in relation to it – and in relation to it as if in relation to a sovereign – that there is below, and above, a right, a left, a forward and a backward, a near and a far.

To my knowledge, this is the only place where the French philosopher directly approached the phenomenological body, which was completely at odds with his thought at that time. But he did not develop these phenomenological reflections further, and the small text has been only recently published.

Foucault also hinted at the phenomenological body, albeit in a critical way, in interviews conducted in Japan in 1978. In connection with the revaluation of body and bodily practices in avant-garde theatre and the martial arts, Foucault did not refer to the embodied dimension of corporeality they imply, and reproduced his own position on the social construction of the disciplined body. In conversations held in a Zen monastery, he referred in a holistic and anti-dualistic manner to meditation as a means of establishing new relations between the body, the mind and the world. But he did not react when master Omori spoke of the universal character of experiences of union between body and mind and body and external world (Foucault 1999: 112–113).

The important use of the notion of experience in Foucault's later work also suggests greater closeness to the phenomenological body (Ortega 1997). The idea of man as an 'animal of experience' (Foucault 1991: 124) points towards the different ways in which the body is lived. The history of the body no longer appears as the history of discourses, but moves towards Barbara Duden's (1997) 'historical somatology'. Similarly, Foucault (1997b, 1997c, 1997d) described certain sexual practices, sadomasochism specifically, as new ways of experiencing the body rather than as discursive productions.

Corporeal action

A fundamental problem with several forms of social constructionism resides in their difficulty in articulating individuals' capacity for agency. Foucault-inspired constructionists tend to avoid the word 'action', to which they prefer the notion of resistance – with its connotation of reacting rather than of displaying agency. This, of course, is not a merely a matter of vocabulary. The choice has anthropological and ontological implications. The terminology of action refers to a view of the human being characterised by the creative agency in context. In contrast, the constructionist terminology of resistance evokes an individual who passively reacts to his environment, and cannot account for non-instrumental relations to the body (Ortega 2007).[6]

In spite of his pronouncement that 'where there is power, there is resistance', Foucault suggested that individuals are constantly subjected to a disciplinary regimen. Yet, as Hans Joas (1996: 169) pointed out, it is necessary to deal not only with 'the emergence of control of the body but also with the development of abilities to ease that control, that is, with intentional reductions in the instrumentalization of the body'. Constructionism resolves the tension between action and structure in favour of the structural, disregarding ways of understanding the body as the basis for action and experience (Howson and Inglis 2001). In contrast, to emphasise the phenomenological body amounts to privileging the body that acts.[7] All action is, in the first place and at the most basic level, bodily action – on that point, the various phenomenological currents coincide.

Phenomenology, with its emphasis on universal bodily structures, may give the impression of essentialising lived experience, and that is one of the reasons why Foucault distrusted it. Such understanding does not correspond to the phenomenological and pragmatic notion of experience as interaction of the individuals with their physical, socio-cultural and historical environments (Shusterman 2012). Moreover, the phenomenological approach does not ignore the structural nor what society does 'to' and 'with' the body. The agents are not dissolved within structures. The latter, however, exist to the extent that agents play a role within them, enacting the structures through their actions, and insofar as the skills and basic dispositions to enact them are transmitted across generations (Crossley 2001: 78). For Merleau-Ponty (2002), such basic dispositions take the form of the body *schema*. Society fashions individuals but at the same time depends on their actions. The notion of *habitus* as used by Pierre Bourdieu articulates body and society, agency and structure: *habitus* arises from the connection between an agent and an already structured domain, but once formed, it contributes to generate and organise new practices, in such a way that there is reciprocity between action and structure. At the same time, showing concern for action is not the same as proclaiming an extreme voluntarism or the primacy of the rational subject. Bodily action transcends the antinomies of free will and determinism, conscious and unconscious agency; but it does approximate, as we shall see next, the 'intentionality without a subject' Foucault referred to in his depiction of power relations.

Corporeal intentionality and power relations

In the end, the phenomenological body may help to explain how power relations are exerted. It is worth recalling that in a 1982 programmatic text, 'The Subject and Power', Foucault (1983: 217) said that what interested him was to find out how power was exercised.

To understand how power relations are exercised from a phenomenological perspective we may in the first place underline that phenomenological and ecological views of corporeality assume the existence of bodily invariants that belong to the 'minimal anthropology' I will examine in the next chapter. These invariants include, among others, humans' erect posture, the asymmetry front-back and the recessive nature of viscerality (Leder 1990; Merleau-Ponty 2002; Todes 2001). Admitting that does not imply a foundationalist or essentialist position, and is compatible with historical and cultural variation. As feminist phenomenologist Iris Marion Young (2005) has stressed, the phenomenological account of basic corporal experiences applies to human existence at a general level, and in no way precludes gender and local specificities.

There is no evidence to conclude that Foucault would have denied the existence of such basic bodily structures. Dreyfus and Rabinow (1983: 111–12), and later Hoy (1999: 6–7) suggest that he would have considered the universality of those structures as too weak to become the basis of criticism and resistance, or to account for the historical specificity of bodily practices. Thus, to speak of pain or pleasure (the latter closer to Foucault's work) as corporeal invariants would be less relevant than to examine how different cultures interpreted them as experiences (Moscoso 2012). Rather than considering bodily commonality among human beings, Foucault was more interested in the limits of that commonality (Levin 2008). Understanding invariants can nonetheless help analyse the exercise of power.

Thus, erect posture or the asymmetry of forward and backward movement are constraints that determine experience. They are constitutive of the human spatial and temporal fields, and are for example responsible for the fact that backward oriented actions are generally disjointed and less effective than forward-oriented ones (Todes 2001). As Dreyfus and Rabinow (1983) note, knowledge of those bodily conditions must have been relevant for the creators of disciplinary apparatuses, as well as for Foucauldian theory, to the extent that it investigates how power manipulates the human body. Indeed, in the first half of the 19th century, European physiology developed an exhaustive inventory of the human body, its functions and invariants, which provided the foundation for fashioning individuals adapted to the productive demands of the era. Physiology increasingly investigated biological laws and causal relations; it ceased to be primarily descriptive and vivisectionist to become a quantitative experimental science, and the paradigm of the experimental method in the life sciences. As discourse of normalisation and disciplining, the science of bodily invariants played a crucial role in the emergence of biopower (Sarasin and Tanner 1998).

The existence of bodily invariants is also significant beyond Foucault's perspective. In a series of well known books and articles philosopher Mark Johnson and linguist George Lakoff have explored how elementary bodily experiences such as balance, force, weakness, self-regulation and the control of physiological processes induce metaphors for speaking about the world and human interaction (Johnson 1987, 1999; Lakoff 1990; Lakoff and Johnson 1980, 1999). The basic schemes of bodily orientation function as preconceptual sense structures and as the foundation for more abstract levels of conceptual orientation. Johnson and Lakoff underline the corporeal foundation of conceptual metaphors, and its persistence within abstract notions. As Johnson (1999: 99) put it: 'We conceptualize and reason the ways we do because of the kinds of bodies we have, the kinds of environments we inhabit, and the symbolic systems we inherit, which are themselves grounded in our embodiment'.

For example, the metaphors that characterise our understanding of moral experience, related to force, light and darkness, discipline and growth among others, emerge out of bodily structures and experiences. Similarly, our notions of mind, memory, knowledge and reason are shaped by corporeal metaphors. Incorporating such evidence into the Foucauldian perspective would clarify the idea that the 'soul' is something disciplinary apparatuses produce inside bodies. If bodily phenomena enter into the making of the moral and cognitive dimensions of mental life, then it makes sense to affirm that different physical regimens 'produce' what Foucault called 'soul'.

Moreover, considering language as an embodied activity also opens the way to countering the claim that, since experience can only be accessed through language, we cannot go beyond language or representation (Butler 2011; Csordas 2001; Weinberg 2012). Such a claim, and its corollary that we can only approach the body through the filter of discourse, hold only if we adopt a representationalist theory of language – one that opposes language to experience, to the body or to reality. If, on the contrary, we assume that language is an embodied activity, then body and language are no longer mutually opposed. Discourse about language itself emerges out of bodily structures. Language does not fashion the body; rather, the body shapes both language and the rational structures we employ to understand the world (Csordas 2001; List 1994; Shilling 2012).

Foucault's assertion that 'power relations can materially penetrate the body in depth, without depending even on the mediation of the subject's own representations' (1980: 186) further illustrates how the phenomenological body may account for the exercise of power relations. These relations penetrate the body without having been consciously or rationally internalised, and thus instantiate what Foucault called 'intentionality without a subject'. Phenomenology adequately describes what he seems to have had in mind. The idea that there is a bodily or motor intentionality highlights the body's capacity to adapt creatively to its environment. There are various terms to refer to corporeal intentionality, such as *body schema* (Merleau-Ponty 2002), *habitus* (Bourdieu 1980, 2010) or *responsiveness* (Todes 2001). They all denote the existence of forms of knowledge and corporeal and practical action that are irreducible to conscious representation or rational choice. Corporeal intentionality elucidates action without appealing to a rationalist or voluntarist framework. It designates a pre-reflexive comprehension that allows us to move and use our surrounding space without having to think about how to proceed. Typing or driving are examples of such practical embodied knowledge. Notions such as *habitus* or *body schema* transcend the dichotomies of free will and determinism, conscious and unconscious agency and even the individual and society (Crossley 2001; Hoy 1999). For Bourdieu (2010: 468), 'the schemes of the habitus, the primary forms of classification, owe their specific efficacy to the fact that they function below the level of consciousness and language, beyond the reach of introspective scrutiny or control by the will'. They are based on pre-reflexive bodily actions, and thus lead to acts that are intentional but not subjective.

We have so far examined central social constructionist claims around human corporeality, as well as the strengths and shortcomings of that perspective when contrasted with the phenomenology of the body. And we have argued that the phenomenological body may help to transcend the dualistic underpinnings of social constructionism, to advance a notion of corporeal action which overcomes the antinomies of action and structure, and finally, to explain how power relations are exercised. Now in the last section of this chapter I would like to address the epistemological and ethical significance of the phenomenological body.

Constructionism and biotechnologies

As already suggested, social constructionism as applied to the body is not inherently subversive or liberating, since it advances a notion of corporeality akin to basic biotechnological understandings of the body, which foster its commodification; conversely, the phenomenological view of the body may provide instruments to critique biotechnological commodification within consumer culture. To make this more explicit let us begin with Ian Hacking's (1999: 6) observation that constructionism tends to assume the following premise: 'X need not have existed, or need not be at all as it is. X, or X as it is at present, is not determined by the nature of things; it is not inevitable.' This premise is often followed by the assertion that 'X is quite bad as it is', and that '[w]e would be much better off if X were done away with, or at least radically transformed'. In contrast, more materialist or phenomenological positions seem essentialist or foundationalist, hence conservative.

There is no doubt that constructionists' historicisation of categories such as race, gender, homosexuality or disability has greatly contributed to denaturalise and depathologise them, and that has certainly constituted an ethical and political step forward for the relevant communities. Social constructionism, however, is not without ethical shortcomings.

To begin with, there is a surprising affinity between constructionist discourses on the one hand and, on the other, certain biomedical and biotechnological breakthroughs. It is often claimed that biotechnological and biomedical possibilities of intervening in our bodies have outdated conceptions of human corporeality as fixed and unalterable, and replaced them by a view of the body as a cultural construction in constant modification. Thus, according to some constructionist thinkers, biotechnologies have the subversive potential to challenge unified, naturalised notions of body and self (Balsamo 1996; Weinberg 2012). The problem, as Llewellyn Negrin has pointed out, is that constructionisms undermine those notions without offering a real alternative. Notions of self and body as fixed and unchanged are, she says, 'stultifying'. Yet 'the advocacy of a constantly mutating self is equally disabling, for without a sense of continuity, it is impossible to act effectively in the world' (Negrin 2008: 93). Moreover, as we shall see below, both constructionism and biotechnologies neglect the materiality and limits of corporeality, and may thereby fuel the commodification of the body (Bordo 1993: 246; Negrin 2008).

The phenomenological approach to the body can sustain a critical stance vis-à-vis constructionist discourses and biotechnological hype without resurrecting the essentialist notion of body that constructionism indeed undermined. The discourse of contemporary biotechnology and technobiomedicine emphasises corporeal malleability (Williams 1997, 2003; Williams and Bendelow 1998). Plastic surgery alters external shape, and tissue can be combined with synthetic substances; bodies become bionic thanks to various devices, such as pacemakers, valves, titanium hips, electronic eyes, cochlear implants, organic and inorganic prostheses, in combination or not with human-machine interfaces (Babo 2004; Clark 2003; Geary 2002). Body parts can be gradually replaced, sometimes with tissue or organs created specifically for transplantation purposes. The commodification of body parts involved in foetal and organ grafting, reproductive technologies and genetic manipulation are a billion-dollar business (Kimbrell 1993). The commercialisation of organs and tissue involves abstracting from subjective experience and objectifying the body so as to use its parts without referring to the individual who is the source (Holland 2001; Sharp 2000). That is why (quite independently of its potential to alter practice) the phenomenologically experienced 'ineradicable presence of the body-as-a-whole' (Leder 1990: 24) constitutes an epistemological and conceptual obstacle to body commodification. In contrast, the constructionist perspective that asserts the fragmentation and malleability of the body seems consistent with it.

Here I am advancing a theoretical and epistemic argument, not a practical or political one, although I believe it also has practical and potentially political implications. It is not because we become aware of our presence as corporeal unified totality that we may concretely oppose body alienation and commodification. My argument is that body commodification underlies a notion of the body as malleable and fragmented. And, as we will see in the next chapter, and as anthropologist Paul Rabinow has emphasised (1996a: 149), it is not the individual, but the fragmented body matter which counts for biotechnological companies. It is in this sense I claim that the body as totality may be an obstacle (at least theoretically) to the commodification of its parts.

Additionally, technobiomedicine, virtual reality, artificial intelligence and transhumanism tend to be seen as sharing a dualist position that, as we have seen, emerges in some versions of body constructionism. The theoreticians of virtual reality and cyberpunk perpetuate a form of mind-body dualism in the cleavage between an immaterial mental self that travels in cyberspace, and a material body condemned to immobility – as staged in Andy and Larry Wachowski's movie trilogy *The Matrix.* In *Neuromancer*, William Gibson's 1984 cult cyberpunk novel, the protagonist Case is expelled from the matrix, where he lived as pure consciousness, free from the bounds of corporeality. 'For Case, who'd lived for the bodiless exultation of cyberspace, it was the Fall. In the bars he'd frequented as a cowboy hotshot, the elite stance involved a certain relaxed contempt for the flesh. Case fell into the prison of his own flesh' (Gibson, 1984: 6). Crafters of cyberspatial utopias, such as the *Society of*

Mind (Minsky 1988) or *The Age of Spiritual Machines* (Kurzweil 2000), imagine the persistence of minds as the condition of a posthuman and postorganic life (Bostrom 2005, 2009; Rubin 2008; Seidel 2009). Cyberspace seems to be a celebration of the spirit, a world in which the subject embodies the fiction of pure thought and is reduced to a sum of information (Halberstam and Livingston 1995; Hayles 1999, 2005; Landow 1997; Le Breton 1999; Plant 1997; Wenner 2002). However, as we will see below, these phantasies of disembodiment downplay the extent to which they presuppose organic and inorganic materiality in the form of circuits, joysticks, plastic and metal components for computer hardware and software, bionic devices, and so forth. Besides, any attempt to overcome or modify or enhance human materiality is necessarily material.

The overrating of mental life and its enhancement at the expense of the material body reproduces the premises of the modern philosophy of subjectivity. We find again here the 'corporeal refusal of corporeality'. David Cronenberg's *oeuvre* provides a cinematic illustration of this phenomenon.[8] In films such as *Shivers* (1975), *The Brood* (1979), *Scanners* (1981), *Videodrome* (1982), *The Fly* (1986), *Crash* (1996) or *eXistenZ* (1999), the body is shown in such a monstrous manner that it provokes abjection (Cruz 2012; Lauretis 2003; Oetjen and Wacker 1993; Riepe 1996, 2002; Riches 2012; Roche 2003; Rodley 1997). Cronenberg's films are obviously far from discarnate. On the contrary, they are *all too* embodied: there are bodies, matter, blood, fluids, viscera throughout. To that extent, they do not advance some sort of dematerialisation or disembodiment. Nonetheless, bodies and matter are treated so as to arouse anxiety, repulsion and abjection. It is therefore not surprising that some critics consider Cronenberg's films as expressions of 'a puritan disgust of the flesh' (Rodley 1997: 157) and of the impossible desire to overcome embodiment. Abjection, writes Elizabeth Grosz (1989: 72), 'is a refusal of the defiling, impure, uncontrollable materiality of the subject's embodied existence'. Cronenberg has succeeded in representing corporeal abjection; his insistence on displaying bodily aberrations, cuts, wounds, fluids, guts, genetic mutations and corporeal metastases gives visual form to the narcissistic re-enchantment of abjection and transforms it into an object of libidinal interest or technological rationality.

Another interesting illustration of the ambivalence vis-à-vis corporeal materiality and the persistence of mind-body dualism is deployed in the work of two of the most prominent cybernetic and cosmetic body performance artists, Sterlac and Orlan. Australian artist Sterlac is generally associated with a claim he has repeated since 1976: 'THE BODY IS OBSOLETE. We are at the end of philosophy and human physiology. Human thought recedes into the human past' (Sterlac 1991: 591). In some performances, the artist's body is suspended with metallic hooks that pierce the skin; in another, he signs his name with three hands – the third one being an implanted robotic organ, which he directs via electric impulses from his abdominal muscles. His performances involve the electronic control of physiology and make use of such processes as electrocardiography, electromyography, the voluntary regulation of pulse and blood flow, as well as sensors for monitoring bodily movement and posture. By means of extremely sophisticated prostheses, Sterlac seeks to demonstrate and overcome the limitations of human anatomy (Appleby 2002; Clarke 2002; Farnell 2000; Goodall 2000; Gray 2002; Jones and Sofia 2002; Jütte 2000; Pitts 2003; Poster 2002; Sterlac 1998, 2000, 2002; Zurbrugg 2000; Zylinska and Hall 2002).

Sterlac's 'Internet Upload' project aims at linking his body to remote sensors in the internet so that his body will be governed by internauts' collective activity rather than through his own agency. Such dissolution of the boundaries of the self represents for him a critique to humanist notions of 'coherence of individuality' and rational agency. The project enables 'multiple choices' for human ontologies, beyond the idea of the self as an autonomous agent (Farnell 2000: 134; see also Zylinska 2002). Sterlac himself (2000: 122) asks: 'Can a body cope with experiences of extreme absence and alien action without becoming overcome by outmoded metaphysical fears and obsessions of individuality and free agency?' He claims that his projects also involve the collapse of power relations. Some critics, on the contrary, argue that they materialise neoliberal notions of choice and individual freedom (Pitts 2003: 163–64; Jones and Sofia 2002: 49).

Sterlac refers to his body as 'the body', and has been criticised for rejecting embodiment (Goodall 2000: 162), as well as for expressing self-hatred and a 'distorted' or 'perverse' Cartesianism (Jones and Sofia 2002: 60). 'The body was split in two', writes the Australian artist about his performance in *Split Body: Voltage-In/Voltage-Out*:

> I was watching my limbs moving in space. I've neither willed that action nor am I contracting my muscles to perform that action. That action is occurring beside, before, it predates myself as a free agent. In other words, half of my body has nothing to do with my free agency.

Sterlac thus personifies the chasm between being a body and having a body, or between *Körper* and *Leib*: the limitations of the lived body must be overcome by the prosthetic enhancement of the physical one.

Notwithstanding their body-denying rhetoric, Sterlac's practices are deeply embodied (Appleby 2002: 101–102; Pitts 2003: 179). Rose Farnell asked him: 'How do you resolve the obvious paradox of having to use your body as the "medium of expression" while simultaneously declaring the body to be "obsolete"?' (Farnell 2000: 133). He answered that it has to do with questioning the status of the physical body and how it is becoming a phantom body (in the sense of phantom limb experiences) through technology (ibid). Sterlac acknowledged having experienced this ambiguity in the flesh, since his efforts to separate the body from gravitational force and surpassing physicality by suspending himself from hooks turned out to be 'brutal' and involved 'lots of trauma' (quoted in Jones and Sofia 2002: 62). As he recognised, 'we cannot operate disembodied' and 'discard the body'; in short, 'You don't escape the body' (quoted in Zylinska and Hall 2002: 121–22).

French Artist Orlan also seeks to undermine notions of a stable and unified identity. Her Carnal Art is characterised as self-portrait in the classical sense, but achieved through technology (http://www.orlan.eu). It is an 'inscription in the flesh', between distortion and reconfiguration, which turns the body into a 'modified ready-made' (Orlan 1989). Orlan undergoes plastic surgeries staged as public performances that involve reading aloud psychoanalytic and literary texts, internet broadcasting, interactive communication, dance and music; they aim at dismantling idealised notions of female beauty and criticising discourses and practices of cosmetic culture (Armstrong 2002; Ayers 2000; Clarke 2000; Jones and Sofia 2002; O'Bryan 2005; Orlan 2002) To the extent that it denaturalises the body and destabilises notions of a unitary stable self, Orlan's work is considered subversive (Jones and Sofia 2002). This view applies especially to her facial plastic surgeries, which challenge Western views of the face as seat of subjectification and identification (Deleuze and Guattari 1987). Orlan too has declared the obsolescence of the body.[9] Her work, she explains, 'is a struggle against the innate, the inexorable, the programmed, Nature, DNA (which is our direct rival as far as artists of representation are concerned), and God!' (quoted in Jeffreys 2005: 163). Such claims target directly the materiality of the body and the inescapable biological limitations that make us up.

Although Orlan's body undergoes radical physical modifications, the material practices to which it is subjected coexist with the artist's view of the body as reducible to discourse. The artist sees herself as working 'on representation' rather than on the body's physicality. For, as she says, Carnal Art 'transforms the body into language, reversing the biblical idea of the word made flesh; the flesh is made word' (Orlan 1989). Thus, in spite of appearances, Orlan's Carnal Art actualises a form of dualism. This is clear in the artist's attitude to pain during her surgical performances. Instead of expressing fear and suffering, she remains 'serene and happy and distant', laughing and playing while her body is being opened up. She claims: 'I don't suffer but I am aware that my body suffers, which are two very different things' (quoted in Ayers 2000: 182, 183). The cheerful body being lacerated can now engage in frivolous and distanced observations: 'Darling, I love your spleen, I love your liver, I adore your pancreas and the line of your femur excites me' (Orlan 1989).

Dualism is also made explicit in an essay by Lacanian psychoanalyst Eugénie Lemoine-Luccioni, which Orlan reads before her performances. The skin 'one has' is therein described as deceptive, in contrast to what one is: 'I never have the skin of what I am . . . I am never what I have' (quoted in

Clarke 2000: 193). Orlan's work turns explicit the lack of fit between identity and image or body and self – a self that does not identify with the body and refuses its corporeal form (Goodall 2000: 157; Jones and Sofia 2002: 58–59). Her Platonism was well grasped by Sterlac when he said: 'I think for Orlan, she speaks of herself as a kind of psyche encapsulated in a physical body' (quoted in Zylinska and Hall 2002: 121).

This is consistent with Orlan's horror of the uncontrollable within the body. She recalls her revolt against breasts and pubic hair growing at puberty 'against [her] will', and the shock of being pregnant and thinking that her 'will alone would cause [her] to abort' (quoted in Jones and Sofia 2002: 60). Such uneasiness reproduces traditional dualist anxieties about the need of the rational self to be fully in voluntary control of the sensuous, affective, carnal and irrational body. The cleavage between the self and the body generates a distanced, objectified and visual relation to corporeality. 'I can observe my own body cut open without suffering!', declares Orlan, 'I can see myself all the way down to my viscera, a new stage of gaze' (Orlan 1989). Such pre-eminently visual apprehension redescribes the body, as we already saw, as an image to be looked at rather than as human flesh to be touched and experienced.

Despite the material interventions in her own flesh, Orlan's Carnal Art wishes to overcome human corporeality through technology. Although both Orlan's and Sterlac's performances deploy a real and continuous engagement with the physical body, their discourse conveys postmodern ideas about bodily alienation and obsolescence. In Sterlac's words: 'IMAGES ARE IMMORTAL, BODIES ARE EPHEMERAL. The body finds it increasingly difficult to match the expectations of its images. In the realm of multiplying and morphing images, the physical body's impotence is apparent' (Sterlac 1998: 123).

The rhetoric of body obsolescence advanced by the theoreticians of cyberspace and posthumanism, as well as by Orlan's and Sterlac's performance art, contradicts their concrete, material and carnal practices, and thus actually highlight the conditions that make us human, the materiality that determines our being in time and space. In the end, as Susan Bordo observes: 'If the body is a metaphor for our locatedness in space and time and thus for the finitude of human perception and knowledge, then the postmodern body is no body at all' (Bordo 1993: 229).

In short, discourses that promise to liberate humans from space-time via virtual reality and biotechnology turn out to be illusory. On the one hand, virtual spaces frequently reproduce the norms that prevail in 'real' life (Balsamo 1993; Negrin 2008; Pitts 2003; Williams 2003). On the other hand, the body, no matter how transformed by surgeries, implants and prostheses, and whether produced by cloning or reproduction, does not disappear. The limitations that materiality imposes can only be surpassed by means of an alternative, cybernetic, cloned, prosthetic materiality. Despite the accompanying rhetoric about body obsolescence, the technologically, surgically and digitally modified bodies of Orlan and Sterlac are still carnal and fleshly bodies. The 'denatured techno-body' is still a material body. In Pat Cadigans's cyberpunk novel *Synners*, notes Anne Balsamo (1999: 280), bodies' materiality is constantly invoked through physical encounters, food intoxication and real death. Even in virtual reality, the body may disappear representationally, but not materially, since it remains engaged with high-tech communication systems (ibid: 284).

Neither does the disembodied techno-body overcome corporeality as source of action. William James's observations remain valid in a postmodern context:

> Where the body is is 'here'; when the body acts is 'now'; what the body touches is 'this'; all other things are 'there' and 'then' and 'that.' These words of emphasized position imply a systematization of things with reference to a focus of action and interest which lies in the body.
>
> (James 1904: 284)

In cyberspace the body does not end – it is redefined. This is enabled by the body's very nature as a live process, as an open structure in constant interaction with the world, as the basis for a 'material self' whose limits, as James saw, are defined by that which the individual considers important for action (James 1890). The flesh is an open system capable of engaging in new possibilities of 'intercorporeality' (Shildrick 2010). Various technologies are changing the form of sensing and understanding the world (Berr 1990; Brooks 2003; Clark 2003; Campbell, F. 2009; Elliot 2003; Geary 2002; Houis, Mieli and Stafford 1999; Roelens and Strauven 2001; Shildrick 2010; Yehya 2001). Devices open up new forms of embodiment; they become incorporated in the body *schema* and expand its range of action. But action itself remains primarily corporeal, so that the body does not cease to be constitutive of action itself (Joas 1996; Merleau-Ponty 2002; Taylor 1989; Todes 2001). Even in virtual reality, Sterlac's assertion that 'THE BODY NOW PERFORMS BEST AS ITS IMAGE' (1998: 123) makes no sense.

Chapter 2

The transparent body

> *The inside, the functional apparatus of the life process, is covered up by an outside which, as far as the life process is concerned, has only one function, namely, to hide and protect it, to prevent its exposure to the light of an appearing world. If this inside were to appear, we would all look alike.*
>
> (Hannah Arendt, *The Life of the Mind*)

The success in our societies of medical visualisation technologies has given unprecedented import to the inside of the human body. Going beyond strictly biomedical areas, these technologies, especially neuroimaging, penetrate the socio-cultural and legal field (Dumit 2004; Ortega and Vidal 2011; Simpson 2012). In popular culture too, medical imaging exerts a unique attraction (Joyce 2005, 2008; van Dijck 2005; van Vall and Zwijnenberg 2009). Even those who have never been subjected to a computerised tomography (CT) scan or a magnetic resonance imaging (MRI) scan are delighted with the medical images shown on television, on the computer screen or in scientific popularisation magazines. Positron emission tomography (PET) colour images of functioning brains have become as popular as Andy Warhol's portraits of Marilyn Monroe or Mao Tse Tung, to which they bear a certain chromatic likeness.

This chapter examines the transformations of corporeality that ensue from visualisation technologies and their popular success. I will explore the hypothesis that the process of making visible the previously invisible interior of the body must be situated within a broader cultural and social transformation: the increased importance given to the body and bodily phenomena in the description and explanation of psychological and subjective processes. In other chapters, I have referred to this transformation using the notions of somatic culture, bioidentities and bioascesis. Nikolas Rose has coined the term 'somatic individuality' (Rose 2007; Rose and Novas 2000) to address the cultural context, in which individuals think, act and speak about themselves and their relations to others drawing on the vocabulary of biomedicine.

My purpose, then, is to inquire into the epistemological, anthropological and socio-cultural consequences of the growing interest in the visualisation of the invisible, beyond its clinical and therapeutic benefits. I will try to show that the visualisation technologies contribute to a disembodiment of subjectivity and, at least in certain contexts, to making the body obsolete. The broad cultural appeal of those technologies may be related to the fact that our physical interior generally escapes subjective and phenomenological apprehension. The objective significance of the inner body does not phenomenologically coincide with subjective experience. Yet insofar as this experience is constitutive of human subjectivity, anatomy's traditional fixation on the corpse, as well as the more recent history of medical visualisation, testify to the weakening of the subjective dimension of corporeality.

The question of the visualisation of the body and the construction of individuality in the context of contemporary medical technologies has been frequently examined (Cartwright 1997; Rose 2007; van Dijck 2005; Waldby 2000). The predominant approaches have been poststructuralist, feminist or variously connected to cultural and science and technology studies. Those perspectives have already contributed to an understanding of the objectification and commodification of the body in advanced liberal societies. I propose to bring into this field the phenomenological approach. By taking a different path towards a subject that has already been extensively analysed, I want to suggest that phenomenology may generate productive insights and ways of critically engaging with how medical visualisation constructs the body. By foregrounding the material and carnal dimension of human corporeality, phenomenology supports a critical stance vis-à-vis the disembodying effect of imaging technologies.

The structure of this chapter is as follows. In the first four short sections I will advance some phenomenological considerations about the ontological status of viscerality and the 'phenomenological anatomy' of the lived body (Leder 1990). The following two sections review some relevant moments in the history of anatomical dissection and the epistemological primacy of the corpse, and the next two sections address epistemological issues around medical visualisation, such as the construction of an objectified body and the rhetoric of transparence. I will then explore the postmodern anatomies of Bodyworlds and the Visible Human Project. The final three sections put forward some general considerations on three interrelated themes that emerge from the phenomenological critique of medical visualisation: the passion for the real, the primacy of vision and the pathos of fragmentation.

Phenomenology of viscerality

In her last book, posthumously published in 1978, philosopher Hannah Arendt (Arendt 1978: 29) makes an interesting observation about the inner body. 'Inside organs', she writes,

> . . . are never pleasing to the eye; once forced into view, they look as though they had been thrown together piecemeal and, unless deformed

> by disease or some peculiar abnormality, they appear alike; not even the various animal species, let alone the individuals, are easy to tell from each other by the mere inspection of the intestines.

Arendt was writing almost a century after Röntgen's discovery of X rays in 1895 launched the interior of the body as an object of major interest – an interest that has remained vivid, not only in the biomedical domain, but also in the broader society, from the law to art to the media. So if the sight of the inner body is not, as Arendt notes, 'pleasing to the eye'; if viscera display a 'monotonous sameness and pervasive ugliness' which contrasts with the 'enormous variety and richness of overt human conduct' (ibid: 35) – whence the interest to make them visible? After all, according to Arendt, 'if this inside were to appear, we would all look alike' (ibid*:* 29).

If Arendt is right in claiming that '"inside we are all alike", unchangeable except at the cost of the very functioning of our inner psychic and bodily organs or conversely, of an intervention undertaken to remove some dysfunction' (Ibid: 37–38), then the drive to visualise the inner body could be interpreted as a way to normalise humans by showing how little their inner bodies diverge. As we shall see in the next chapter, in the context of bioascesis and the quest for perfect bodies, individuals sometimes indeed try to conform to norms about bodily appearance in ways that are consistent with such an interpretation. We have inner body normalisation in the case of visualisation technologies, and bodily appearance normalisation in the case of bioascesis. But Arendt's claims must be nuanced. Gunther von Hagens's 'anatomic art' of plastinated corpses, or Orlan's 'carnal art' performances, in which the artist undergoes plastic surgeries transmitted live (Featherstone 2000; O'Bryan 2005; Pitts 2003) suggest that the inner body is being domesticated as a site of subjectivity. Once the inner body gains the visibility it has in our culture, then it becomes the object of attemps at giving it marks of singularity, at integrating it fully into individual identity.

Phenomenological anatomy of the lived body

In *The Absent Body*, Drew Leder proposes a phenomenological anatomy that is of fundamental relevance for our argument. Contrary to most phenomenologists of corporeality, who begin their analysis with what Merleau-Ponty (1964) called the 'primacy of perception', Leder starts with the absent and recessive dimensions of the lived body – from the self-effacement of bodily organs such as the eye, which are the origin of the perceptual field, to the experiential disappearance of visceral processes, which take place almost automatically beyond conscious perception. Corporeal presence, Leder argues, is paradoxical because it simultaneously involves an inexorable presence and a fundamental absence, organs and functions that appear in the foreground of experience, while others remain in an invisible background. According to

Leder (1990: 3), such a situation sustains dualism – which is therefore not, as we usually believe, a metaphysics or an ontology foreign to lived experience. Hannah Arendt (1978: 22–23) maintained a similar view when she wrote that 'the two-world theory belongs among the metaphysical fallacies but it would never have been able to survive for so many centuries if it had not so plausibly corresponded to some basic experiences'. In short, questioning the conceptual hegemony of dualism is not incompatible with taking into account its experiential truth.

Insofar as it is the starting point of our existence in the world, the 'zero point of all perception',[1] the surface of the body tends to disappear from thematic attention; projected towards the outside, the organs of perception and movement are transparent while they are being used. Generally when I perceive through any given organ, the organ itself recedes from the perceptual field it discloses: I do not see my eye, smell my nasal tissue or hear my ear, although I perceive through all these organs (Leder 1990: 14–15). In contrast with this superficial bodily dimension, which extends towards the outside via action – what Leder calls 'ecstatic body' – viscerality remains invisible as the part of the body 'which we do *not* use to perceive or act upon the world in a direct sense' (ibid: 53). Leder calls such recoil of viscerality 'deep disappearance' (ibid). 'Deep' must be understood in a double sense: physical, since the interior of the body is protected by the body's own outer surface; phenomenological, to the extent that it is not available to conscious apprehension and control. From the point of view of Leder's analysis, the interoceptive field, that is, the sensations of the inner organs, possesses three main features. First, it is qualitatively reduced relative to the exteroceptive field. It lacks the multidimensionality of the external senses; and since it possesses a much smaller number and variety of sensory receptors, its range of response is more limited. Secondly, visceral sensations lack the spatial precision of exteroception, the kind of accurate localisation displayed for example by skin sensations. Finally, the flux of interoceptive experience lacks the spatio-temporal continuity of external sensations: most vegetative processes take place silently.[2] Thus, '[u]nlike the completed perception of the proprioceptive body, our inner body is marked by regional gaps, organs that although crucial for sustaining life, cannot be somesthetically perceived' (Leder 1990: 43).

Minimal anthropology

The body's recessive nature is one of the existential and biological invariants that shape human experience (Leder 1990: 150). These invariants belong to the 'minimal anthropology' mentioned in the previous chapter. The upright position, which must be constantly maintained within the earth's gravitational field, as well as the asymmetry of our motion, orientated towards the front and the back, are such invariable features. Backward movement is generally disjointed and less effective than forward motion. Hence, action and active perception are always oriented forward. When we move, observes philosopher Samuel Todes (2001: 49), 'we do not merely notice but *produce* the spatiotemporal field around us' (see also Hoffman 2001).

Cultural variations, Drew Leder (1990: 3) emphasises, 'are always played out upon the keyboard of possibilities presented by our corporeal structures'. These structures define minimal anthropology, but biotechnological developments in areas such as genetic engeneering, prosthetics, nanotechnology, robotics, synthetic biology or neurotechnology may modify them (Broadhurst and Machon 2009; Clark 2003; Cleland 2010; Shildrick 2010; Smith and Morra 2006; Zylinska 2002). It is not impossible to imagine a not-so-distant future in which what we now consider existential-biological invariants will be different. Medical visualisation technologies and the emphasis on making visible the inside of the body will perhaps affect the recessive tendency of viscerality. Such alterations may give rise to new anthropological configurations that will be modulated by culture.

To be able to talk about variation we always need an invariant by reference to which the variation appears as such; it is necessary to assume the constancy of the human body. For the active person, Todes observes (2001: 71), the body is 'the condition of the possibility of things appearing variably to him', the fixed reference. Thus, the '[a]wareness of our *own* movement implies a pervasive and systematic change of *all* circumstantial positions in respect to our own. . . . By moving, we transform our field of perceptual positions' (ibid: 108). The moving body carries with itself its point of origin. No matter in which direction one moves, the body remains the absolute 'here' around which the different 'theres' are organised. When I am active, I can turn around and reverse the direction of my movement; and yet, the sense of direction remains constant, even when what used to be before me is now behind. This follows from the functional assymmetry of our front-back body structure.

Admitting the existence of a 'minimal anthropology' does not imply an essentialist downplaying of cultural and socio-historical variation. But it does involve taking into account what Drew Leder (1990: 150) calls 'phenomenological vectors', structures of experience which predispose the subject to certain types of action. For example, in all known cultures, humans use their hands to work and explore the world; this is enabled by the hand's plasticity, including the dexterity of the thumb, as well as the freedom made possible by humans' upright position. The important point is that such phenomenological vectors acquire specific forms from particular socio-cultural contexts, and can therefore not be conceived independently of their historical and cultural embodiments. Other forms of cultural modulation relate to the social construction of the consciousness of the body. They reside in what Leder calls 'social dys-appearance' – the appearance of the body in conditions that are considered bad, abnormal or impaired (1990: 84). Here, the body is somehow distanced from itself by the incorporated gaze of the other, and that brings about its explicit thematisation (Ibid: 96). Franz Fanon's description of the black man as locked in his body offers a good illustration of social dys-appearance. As Burwood (2008: 270) explains:

> When Fanon encountered the look of the white other, his experience of objectification was thus amplified by these images. He describes himself being fixed by the glances, movements and attitudes of the other; of being burdened by an unfamiliar weight; of the real world challenging his claims; of his bodily schema crumbling, to be replaced by a racial epidermal schema or a historico-racial schema; of being made aware of his body in an unfamiliar and detached way; of occupying space; of being made responsible for his body, his race, and his ancestors; and of being battered down by the imagery of racist stereotyping.

In short, phenomenological vectors and forms of social dys-appearance coexist with biological-existential invariants such as the recessive character of the interior of the body.

If Merleau-Ponty, Husserl, Leder, Erwin Strauss and others are right to say that the motor intentionality of the body corresponds to the formula 'I can' (walk, look, learn and so forth), then the formula for the inside of the body would be 'it can'. The neutral pronoun 'it' refers indifferently to any of the interior organs of the body, since those organs function automatically, and are absent from our perception, will and action. Nevertheless, the interior's 'it can' operates as an 'I must'. It defines a corporeal dimension which escapes my awareness and over which I have no control – and yet which impels me and compels me. Even here, however, humans can learn to affect viscerality, as in the Eastern practices of attention that lead to a grasp of consciousness of functions such as pulse and heartbeat, and subsequently to their voluntary control. Empirical studies have indeed demonstrated a strengthening of parasympathetic (vagal) heart rate control in regular yoga practitioners (Sunkaria, Kumar and Chandra Saxena 2010). The extreme level of training required to attain such bodily control is a confirmation, rather than a refutation, of the recessive nature of the interior of the body.

One objection to my argument could be the existence of what anthropologist Margaret Lock (1993) names 'local biologies'. The term refers to the processes whereby the interactions of social and biological factors result in biological differences. Drawing on her anthropological study of menopause and the existence of variations in symptom reporting, Lock came to reject the biomedical view that conceptualises menopause as an 'invariant biological transformation' subjected only to superficial modification by socio-cultural and psychological factors (Lock and Nguyen 2010: 89). Local biologies, however, also emerge in connection with the limited number of possibilities enabled by our corporeality – as Lock and Vinh-Kim Nguyen acknowledge when they observe that the experience of the end of menstruation 'is undeniably "real", and appears to be biologically programmed to take place in the majority of women between their late forties and mid-fifties' (ibid).

The ontological ambiguity of viscerality

The interior organs are characterised by an 'ontological ambiguity' (Young 1997: 101). They seem to belong to me, but I also belong to them insofar as they are indispensable for my existence. They constitute a 'foreign-mineness', to use Frederick Buytendijk's expression (Leder 1990: 48); and precisely insofar as I am alienated from them, the visceral 'cannot be properly said to belong to the subject; it is a power that traverses me, granting me life in ways I have never fully willed nor comprehended' (Leder 1990: 65). Understanding the experience of the lived body requires combining a third-person scientific perspective with a first-person phenomenological description. I experience my body subjectively but also as an object among other objects, as *Körper* and *Leib*, as physical body and lived body (see Chapter 1). Viscerality embodies a third-person dimension (the body-object or physical body) to which I have no immediate access in first-person experience (the body-subject or lived body). It represents an 'alienness-from-within': were they presented to me from the outside, I could not recognise my internal organs as mine. 'I would stand in the presence of something uncanny and alien – could such organs be inside me?' (Leder 1990: 54). The interior of the body, then, is *unheimlich* in Freud's (1919: 268) sense – something uncanny, yet at the same time secretly familiar.

If the interior of the body is indeed experienced as 'secretly familiar', then we could hypothesise that the fascination of visualisation technologies is due less to their diagnostic and therapeutic efficacy, than to the fact that they make sense of, and give sense to the constitutive ambiguity of the visceral, and thus transform it into something strangely familiar. Visualisation technologies may thus provide access to the interior of the body, and enable or give the illusion of a certain control over its phenomenological dimension. Hence the contrast this chapter emphasises between viscerality as a non-objectifiable corporeal dimension on the one hand, and on the other hand the biomedical objectification of the lived body, together with the resulting disincarnation of subjectivity which is part and parcel of the history of visualisation technologies. In the perspective of that history, viscerality is the remainder, the *restance* in Derrida's (1992) word – that which resists the objectivation of the body. Visualisation offers only an indirect apprehension of the interior body, without involving subjective experience, so that 'the absences that haunt my bodily depths are not effaced by these reflective maneuvers'.[3] The 'phenomenological noncoincidence' (Leder 1990: 44) between the body-object and the body-subject remains. Emphasising the phenomenological primacy of viscerality is a way of reappraising the body as the subject of experience and action.

The development of visualisation technologies may be linked to the convergence, in Western culture, between knowledge of the self and knowledge of the interior body. During the 16th and 17th centuries – the 'visceral century' according to Hillman (1997: 83) – the interior of the body was involved in the generation of mental and spiritual interiority, of private individual experience. Opening the body was crucial for the production of knowledge (Hillman 1996, 1997; Sawday 1996). Realising the Delphic precept 'Know thyself' required anatomical knowledge. In a radically different framework, the fascination of contemporary visualisation technologies manifests the continuity of the conviction that personal identity is inseparable from materiality. This is further attested by the continued relevance of medicine for thinking about the human. In *The Birth of the Clinic*, Foucault (1994b: 245) remarks: 'The importance of Bichat, Jackson, and Freud in European culture does not prove that they were philosophers as well as doctors, but that, in this culture, medical thought is fully engaged in the philosophical status of man'. By introducing death into medical thought, pathological anatomy made possible the emergence of the modern subject, whose status as an object of knowledge starts with the fundamental experience of finitude. In the West, modern medicine and philosophy share a fixation on death and on the choice of the corpse as favoured model of the body. As we explore in this chapter, in the anatomical tradition, in imaging technologies and in philosophy, the corpse supplants the living body as model and has therefore epistemological primacy.

We shall begin this exploration by reviewing some crucial moments in the history of anatomy and visualisation. Our goal is not to write a history of dissection or of medical imaging. Excellent research exists on both topics (for the former, see Carlino 1999; Mandressi 2003; Park 2010; for the latter, Burri and Dumit 2008; Cartwright 1997; Dijck 2005; Dumit 2004; Joyce 2008; Kevles 1997; Stafford 1996; van de Vall and Zwijnenberg 2009; Waldby 2000). Rather, we shall bring together elements that illuminate how medical dissection and the technologies of body visualisation have implied objectification and disembodiment.

Open up a few corpses

Western biomedical rationality is characterised by the dissection of corpses as a form of production of knowledge. Anatomical dissection was not the main source of knowledge of the body in other traditions, such as Chinese, *Ayurveda* Indian or *Unani* Arab medicines; nor did it play that role in archaic Greece, which ignored the idea of a discrete 'organic' body. For the latter, individuals did not possess a notion of body as unity, organised structure. Instead of the body as a whole they referred to its parts, members and articulations (Snell 1953). Owing to the lack of distinctive elements for bodily individualisation, the somatic dimension was linked to social relations. The centre of the person drifted to the genealogy or to the individual's powers and attributes. It was therefore impossible in such a context clearly to delimit individual and collective illness (DeHart 2000: 357–58).

Unlike the Homeric tradition, the Hippocratic schools regard the body (*soma*) as a unified whole. The physician learned about the body through the science of regimen, dietetics (which applied in the case of internal illnesses) and through surgical practice, which was reserved for handling external wounds. On the one hand, the surgical body remained a surface body; the practical orientation of surgery obstructed the development of an inner anatomy that would be guided by visual observation. Since anatomical knowledge responded largely to the functional demands of how to best fix bone injuries, the surgical approach did not generate specific morphological knowledge (DeHart 2000).

On the other hand, Hippocratic medicine excluded anatomy by virtue of its extreme concern for the dignity of the body (Annoni and Barras 1993; DeHart 2000). In the Hippocratic oath, the physician committed himself not to 'cut, even for the stone', but to 'leave such procedures to the practitioners of that craft' (Hippocrates 1983: 67), namely to the surgeons, for surgery was considered an inferior manual profession. Such an attitude changed around the fourth century BCE under the influence of Platonism: once the soul is radically dissociated from its corporeal envelope, the corpse no longer has anything to do with the person; as a material object without feelings or rights, it can be freely anatomised.

Such a change in perception opened the way for the first human dissections, practised in Alexandria in the third century BCE. Up to that point, human dissection was not thinkable. Human internal morphology was deduced from the body's external form or by analogy from the internal morphology of dissected or vivisected animals (Carlino 1999; Mandressi 2003). Most important is the fact that, starting with the texts that composed the *Corpus Hippocraticum* of the fifth century BCE and until the 17th century, humoral theory remained the dominant medical paradigm. Within this paradigm, health was understood as resulting from the balance of the four basic humours (blood, phlegm, yellow bile and black bile) and their basic qualities (hot, cold, wet and dry); this balance could be altered due to internal or external causes, and healing amounted to its restoration (Carlino 1999; Mandressi 2003; Porter 1999; Zaner 1992). As long as humoral theory sufficed for etiology and diagnosis, anatomy proved unnecessary. However, insofar as it helped locate the humours within the body or follow their movements, there was not inherent contradiction between the humoral paradigm and anatomical knowledge.

Excluding a short interlude in Hellenistic Alexandria in the third century BC, during which the practice of anatomy gained unprecedented relevance and included the dissection of corpses and, possibly, the vivisection of slaves,[4] anatomy remained ruled by the Galenic paradigm and excluded human dissection until the 13th century (Carlino 1999; Mandressi 2003). For

Galen, anatomical knowledge, together with logic, helped to formulate proper diagnostics and therapeutics. However, that knowledge was clinically inadequate since it remained difficult to reconcile with humoral pathology. Thus, anatomy, wrote Galen in the *De anatomicis administrationibus*, was 'more useful for philosophers than for physicians' (quoted in Carlino 1999: 4).

Galen nonetheless performed numerous dissections and vivisections on animals, especially monkeys, whose internal apparatus and morphology most ressembled the human. His extrapolation to human anatomy was to be later criticised in the tradition of Vesalius (Carlino 1999: 143–44). Galen insisted on observing both animal and human bodies. However, the observation of the body was for him an auxiliary tool for teaching the medical disciplines which were largely constituted around learning texts (ibid: 149–50). For about a millenium, dissections, or more precisely, prosections, were associated with the reading and commentary of canonical medical texts. Their purpose was to demonstrate the texts, and they were of little use outside that didactic context. The image of the anatomised body reinforced the instructor's words, illustrating and corroborating the contents read, which were usually Galen. A beautiful illustration of the dissociation between the instructor, the *lector* who reads and the barber or surgeon, and the *sector* who carries out the dissection, is provided by an image from the Italian translation of Johannes de Ketham's *Fasciculo di Medicina*, of 1493. The image depicts an anatomy lesson by Mondino (Fig. 2.1). The sector, holding the scalpel, is the only character who does not wear a robe; this emphasises the non-academic and manual character of dissection. While the *lector* reads passages from classical anatomical texts, a third character, the *demonstrator*, indicates with a pointer the areas the *sector* should incise, according to what is being read or recited.

According to Carlino (ibid: 156), two obstacles had to be overcome for dissection to be practised again in the West and for anatomy to recover its dignity. First, it was necessary to react against the principle of *auctoritas* and admit that Galen's descriptions might be wrong. Secondly, it was necessary to surmount the horror traditionally linked to the handling and desecration of corpses. 'For no one', writes Aristotle in *De partibus animalium* (1995: 645a 29–30), 'can look at the elements of the human frame – blood, flesh, bones, vessels, and the like – without much repugnance'. This repugnance, however, did not imply a prohibition or an insuperable barrier to anatomical knowledge. Explaining the neglect of dissection for a millenium requires taking into account various causes, both external and internal, to the history of medicine. Considering that none is by itself satisfactory, historian of medicine Rafael Mandressi (2003: 60) inverts the perspective and looks at the emergence of dissection as an 'epistemological event'. Why, he asks, did dissection become necessary to the study of the human body? Dissection was adopted as a practice at a time when it was perceived as useful to the knowledge of the human body (Ibid: 61). One of the main factors which influenced this

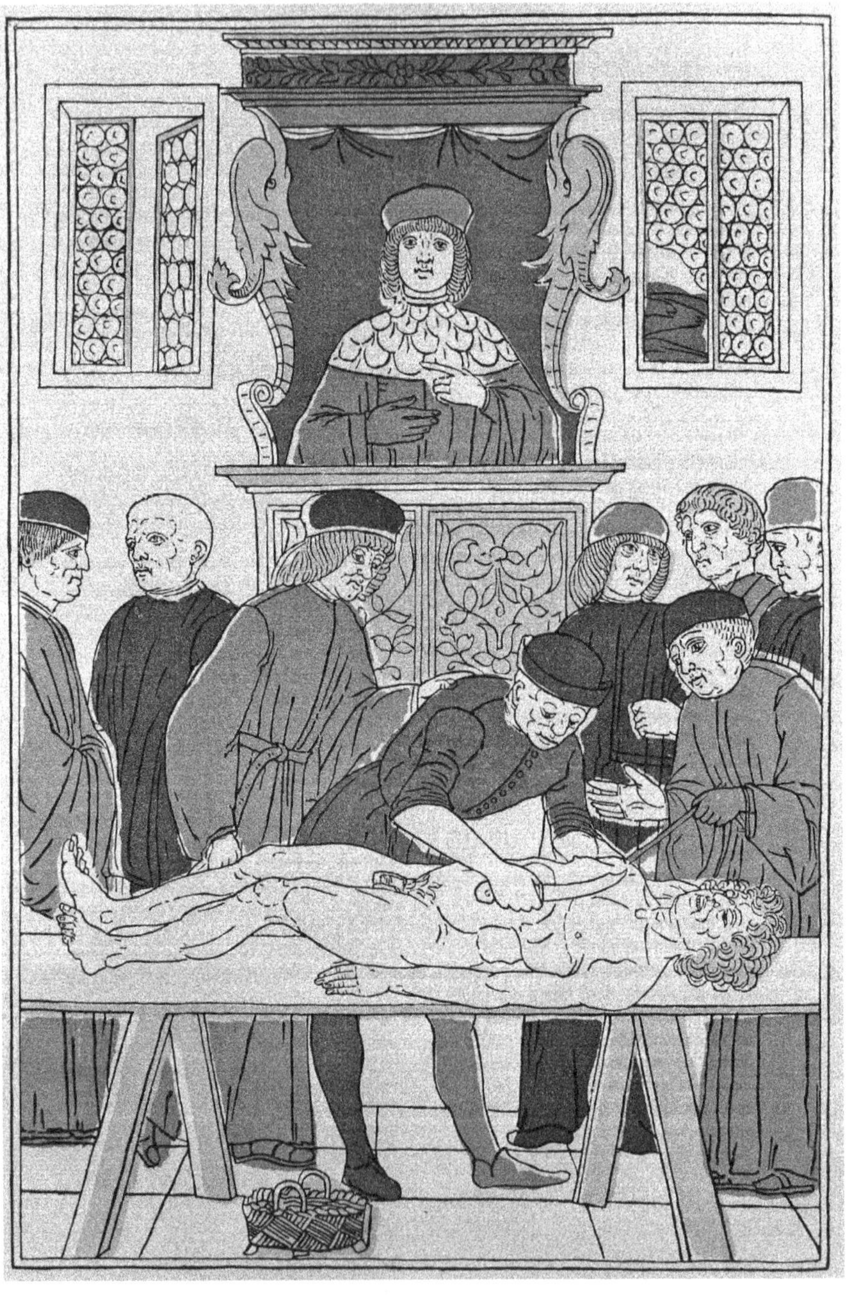

Figure 2.1 Johannes de Ketham, The *Fasciculo di Medicina*, Venice 1493, Woodcut of anatomical dissection

Credit: Wellcome Library, London

epistemological change was the translation into Latin, between the 11th and the 14th centuries, of ancient scientific and medical Greek texts that had been transmitted in Arab versions. These works emphasised visual observation and promoted anatomy as the first branch of the sciences of the body. In addition, in the 12th and 13th centuries, dissection was introduced in funerary and juridical contexts.

Historians agree that the first unequivocal narrative of a dissection carried out for the purposes of anatomical study places it in Bologna in 1316 and attributes it to Mondino de' Luzzi.[5] However, as Katharine Park shows in her book *Secrets of Women* (2010), starting in mid-13th century Italy, autopsies were commonly carried out to determine the causes of death. The dead bodies of holy and criminal people were also opened in the search for material signs of sanctity or diabolical presence – signs such as a cross or the instruments of the passion in the former, a hairy heart or an extra rib in the latter (Park 1994, 2010). Park notes that historians of medicine have usually focused on public dissection as a teaching and scientific resource. They have therefore paid comparatively little attention to dissection practices that were unrelated to medical practice, such as autopsies carried out in a legal context or in private settings, in connection with the cult of relics, foetal excisions or funerary rituals (Park 2010: 14–16). The emphasis on saints' relics and funerary cults demonstrates the religious significance of dissection, and contradicts a common view among historians of anatomy (eg, Pouchelle 1983), who contrast medieval religiosity to the scientific rationalism of the Renaissance and invoke the Church's reservations vis-à-vis opening the dead body (Park 2010: 21).

Carlino highlights three elements that contributed to establish dissection as a legitimate practice in anatomical research and education in the 14th century: the circulation of classical Greek and Arab texts, the development of university teaching and the already-mentioned practice of autopsy (Carlino 1999; Mandressi 2003). The recourse to classical sources and the appeal to the authority of antiquity manifested the usefulness of anatomy for medicine and natural philosophy, and gave these disciplines epistemological and anthropological legitimacy. At the same time, in spite of the renewed value of the anatomised body, it inhibited the development of dissection practices.

The earliest documented dissections carried out for anatomical study retain an eminently didactic character. Mondino remained true to the Galenic paradigm and did not question Galen's anatomical descriptions. The goal of dissection was still to demonstrate the written text, rather than empirically to explore the inner body. The image of the anatomised corpse reinforced the teacher's word, which itself reproduced the canonical text the students had to learn. Only two centuries later, with Vesalius, would the authority of the ancients be replaced by the direct evidence provided by the observation of the dissected corpse.

Vesalius's *De humani corporis fabrica*, of 1543, has been traditionally regarded as a hallmark in the history of anatomic dissection. In this work the anatomist questioned Galen's authority and criticised his followers for adhering completely to the master's word and refusing to correct his descriptions. Vesalius inverted the hierarchy between textual authority and empirical evidence and gave new dignity to the latter. He thereby rehabilitated surgery, which had been despised as a manual craft, and characterised it as 'the foremost and most ancient limb of medicine' (Vesalius 1998 [1543]: xlix). The traditional dissociation between physicians and surgeons seemed to him an 'evil fragmentation' of the art of healing (ibid: li), a false dichotomy that Vesalius himself surmounted in his own approach to anatomy. The double, theoretical and practical, function of anatomy is patent in the frontispiece of *De humani corporis fabrica*, which exhibits Vesalius, surrounded by a vast audience and dissecting the uterus of a naked woman (Fig. 2.2). As an 'iconic manifesto of the new anatomy' (Carlino 1999: 44), this image would for over a century be the iconographic model of the public anatomy lesson. With Vesalius, dissection becomes the main means of revealing the secrets of the body without the mediation of a text (Park 2010: 207). The new didactic and investigative approach counters the earlier subordination of observation to description. Henceforth, 'the body dictated the text, with an emphasis on the decisive contribution of the visual to both anatomical teaching and research' (Carlino 1999: 1).

It would nevertheless be mistaken to see in the *Fabrica* a symbol of anti-Galenism or a radical rupture with *auctoritas* in favour of direct observation of the dead body. Vesalius was certainly not the first to question Galenic assumptions, and the anatomical science that was being constituted in his time developed an active relationship with Galen's work, reinterpreting and adapting it (Mandressi 2003: 82–83). As Katharine Park has noted, the *Fabrica* embodies Vesalius's conscious attempt at fashioning himself as a scientific icon, a transgressor and reformer of anatomical science endowed with epistemic authority thanks to his commitment to uncover the truths hidden inside the human body. In the history of medicine, she writes, Vesalius 'positioned himself as both the restorer of the lost empire of medicine and the founder of a new dynasty of great physicians, on the model of Apollo's establishment of the Asclepian line' (Park 2010: 247). The authoritative book, as the frontispiece shows, is no longer placed in the teacher's hands, but in those of a student oblivious of the grand scene that surrounds him. An older man to the student's right points towards Vesalius and the dissection table, as if calling attention to the epistemic centre of an event whose ultimate significance is epitomised in the simultaneous presence and the spatial disposition of the uterus and the skeleton. Through these associations with the drama of life and death, Vesalius highlighted his role as founder of a new anatomical science based on the empirical observation of the human corpse (Carlino 1999; Mandressi 2003; Park 2010; Sawday 1996).

Figure 2.2 Andreas Vesalius, *De humani corporis fabrica libri septem*, 1555, Title page

Credit: Wellcome Library, London

As also illustrated in the frontispiece of the *Fabrica*, the public anatomy lesson was a public ceremony that attracted a large audience, made up of professors, students, physicians, artists and a general public, for which it constituted a sophisticated form of entertainment. In Italy, Holland and other European countries, anatomical theatres were constructed, with their galleries organised, as in Padua, according to social rank. Such theatres added to the prestige of the city where they were built (Ferrari 1987; Klestinec 2004, 2007, 2011; Park 1994, 2010; Tierney 1998). As conspicous in Vesalius's frontispiece, as well as in an image of the Leiden theatre (Fig. 2.3) in which skeletons display the well-known aphorisms *Nosce te Ipsum* (Know Thyself) and *Nascentes Morimur* (we are born to die), public dissections conveyed a moral lesson. Anatomical theatres functioned as an 'extravagant architectural lesson in human mortality' (Sawday 1996: 72); they were 'moral tales' in themselves and 'amphiteaters of moralized anatomy' (Lunsingh Scheurleer 1975; see also Cook 2006; Sawday 1996; Turner 2002: 205–206).

Starting with Vesalius, for Western medicine, the truth about the body and illness is to be found inside the body, in the visualisation of the invisible. The emphasis on visuality gives access to the secrets of the body without the

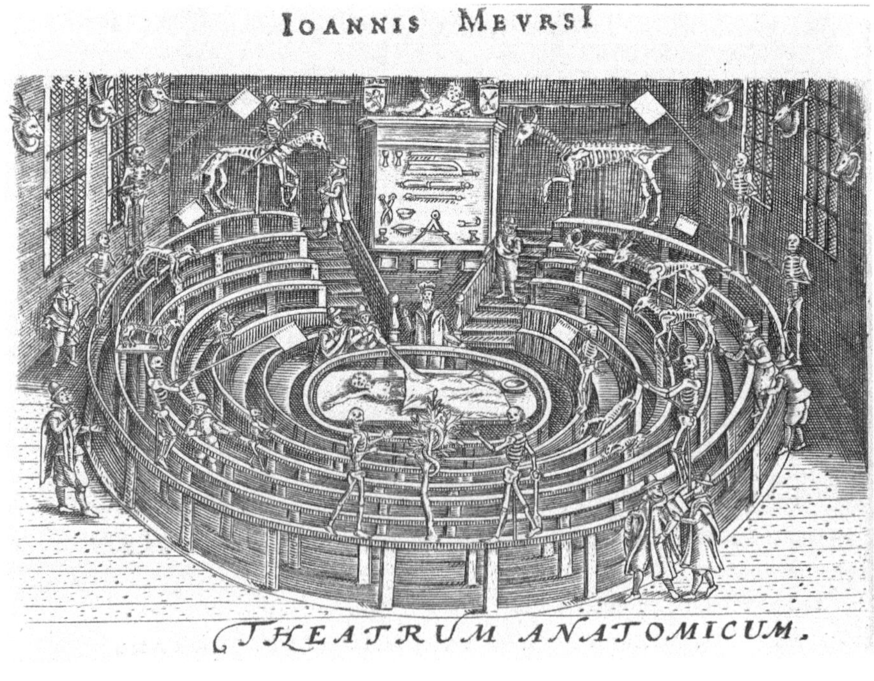

Figure 2.3 Johannes Meursius, Engraving of an anatomical dissection at Leiden, c. 1614
Credit: Wellcome Library, London

mediation of the text since, as mentioned; the body itself produces henceforth the text. In Jean Starobinski's (1985: 162) elegant words:

> The complexity of the nervous system (of which the ancients knew nothing) was revealed by the fine, cold tip of the scalpel for all to see, and then preserved by the burin of the engraver upon the pages of a printed book.

In *De humani corporis fabrica*, the image becomes a 'translation medium for the text' (Bergmann 1996: 89). The production of images replaces words in the construction of truth. Thus emerges the close relationship between visuality and the scientific knowledge of the body; the model of an infinitely reproducible body becomes the norm for all bodies. And although this model is based on the dissection of corpses, Vesalius's prints or Leonardo da Vinci's drawings do not seem to be of parts of dead bodies, but place the spectator before a 'visionary animated anatomy' (ibid: 91). Death, in the new medical gaze, thus becomes the model of life, and the corpse the paradigm of the living body.

In phenomenological terms, the reduction of the experience of the subjective body to the knowledge of an objective, measurable, quantified and fragmented body, which since the Vesalian revolution accompanies anatomical practices and visualisation technologies, corresponds to the subject's relation to his or her body as something he or she owns, rather than that which he or she *is*. The vocabulary of 'possessive individualism' (Macpherson 1962) permeates the genesis of the modern body, and defines property relationships as a constitutive part of human nature, and as a condition for realising human potentiality. The individual is free as owner of his or her body (Duden 1997). To *have* a body becomes more important than to *be* a body; the body-thing, with its physical properties, obliterates the body *I am*, the physical and lived body. The reversibility of the physical and the lived body (Merleau-Ponty 2002), that is, its capacity to be simultanously sentient and sensible, as when one presses two hands together, is, as we saw in the previous chapter, an ontological property of human corporeality, not a historically contingent one. It would be naive to claim that before the emergence of possessive individualism and the anatomical revolution of the Renaissance humans were *Leib*, while those that came after merely have *Körper*. The Western anatomical tradition and the history of the visualisation of the body nonetheless attest to a significant weakening of the latter compared with a model patterned after the body-machine, the body-corpse or the virtual body.

The epistemological primacy of the corpse

'[W]hen I am dead', Socrates says in Plato's *Phaedo* (2005: 95), 'I shall not stay, but depart and be gone'. He thus calls attention to the particular temporality of one's own corpse, and inaugurates a tradition of first-person reflections about it. In Antiquity and the Middle Ages, such reflections belonged to a personal, existential experience of finitude. With Descartes, the corpse ceased to be apprehended in the first person and became a mere object of dissection. In the thought experiment of the second *Meditation*, he wrote: 'I consider myself as having a face, hands, arms, and all that system of members composed of bones and flesh *as seen in a corpse* which I designated by the name of body' (1647: 64, emphasis added).[6]

While Socrates referred to his own corpse in an anticipatory manner, Descartes alluded to someone else's, and took it as point of departure for thinking about his own body. Now, a corpse seen in a third-person perspective is the phenomenological opposite of that seen in the first-person. 'It lies there', writes Leder (1990: 146), 'strangely unmoving, unseeing flesh, no longer a play of absence and reference'. The third-person view of the corpse shapes the understanding of the body in modern thought, and specifically in biomedicine. It provides the paradigm for the lived body; at the same time, as a methodological tool and a regulatory ideal, it hinders other forms of approaching human embodiment. Medicine furnishes ideal and normative models for the body. From its public representations in early modern anatomical amphitheatres to contemporary imaging techniques, the body of an anonymous other is offered as *the* body. We therefore reconstruct our own bodies on the basis of models created with corpses, looked at in a third-person perspective and devoid of subjective dimension. In death 'all the linkages of the lived body are disrupted. The body dissolves from an operational whole into discrete organs and tissues, which can be studied in isolation. This then fosters an atomistic concept of disease and therapeutic response' (Leder 1990: 147).

The epistemological primacy of the corpse in medical thought sustained the growth of medical technologies, especially visualisation techniques in which access to bodily tissue does not require the individual's death. Thus the doctor who in 1835 commented on the stethoscope: 'we anatomize by auscultation (if I may say so), while the patient is yet alive' (quoted in Reiser 1990: 30). The patient is reduced to a body to be treated at the expense of the subjective and psychosocial determinants of illness – an attitude captured in the title of a 1985 article by Richard Baron, 'I Can't Hear You While I'm Listening' (Leder 1992: 22). Although such reductionistic model of biomedical rationality coexists with other investigative and therapeutic approaches, the body as objectified and fragmented continues to furnish normative models. Visualisation techniques allow us to apprehend our bodies only as an image, and that necessarily distances it from corporeal experience. Noting that his reaction to his own image was 'But I never looked like that!', Roland Barthes replied (to himself):

> How do you know? What is the 'you' you might or might not look like? Where do you find it – by which morphological or expressive calibration?

> Where is your authentic body? *You are the only one who can never see your self except as an image; . . . even and especially for your own body, you are condemned to the repertoire of its images.*
>
> (*Roland Barthes by Roland Barthes* quoted in Sawday, 1996: 15, emphasis added)

Barthes's questions summarise the situation in which visualisation technologies leave corporeality. The visualised body has lost its subjective dimension and persists only as body-object, as *Körper* rather than *Leib*. We see the body we have, not the body we are; the image represents the body in the third, not in the first person, and thus opens the way for an external comprehension of interiority.

According to historian Rafael Mandressi, it is this potential for disclosing human interiority that turned anatomy into a powerful metaphor to describe human introspection in 17th-century French Christian spirituality. The 'interior reform' advanced by the theologian Jean-Pierre Camus consisted in dissecting the human soul in the way anatomists dissected the body. Thus (quoted in Mandressi 2003: 234), he writes:

> in the same way that before undertaking to cure human bodies, physicians diligently study anatomy and examine in detail the body's composition, if we wish to reform us innerly and put back on the right track the soul that has diverted from its duty, we must see very clearly in all its springs, and penetrate its folds, turnings and hiding-places.

I would like to suggest that the powerful analogy between the physician who studies anatomy before endeavoring to cure and the Christian who examines the secret corners of his soul in order to reform his inner life is more than an analogy, and that it points to the ways in which the interior of the body relates to psychological or spiritual interiority in the Western philosophical and anatomical traditions. As we shall see next, this relationship between corporeal and psychological interiority extends to the technologies of medical visualisation.

My God, I see!

The title of this section is taken from a chapter in Thomas Mann's *The Magic Mountain*. Entitled 'Sudden Enlightenment', the chapter describes the protagonist Hans Castorp's encounter with X-rays. Castorp, a young engineer from Hamburg, goes to visit his cousin in the Swiss sanatorium Berghof, where he ends up staying seven years instead of the intended three weeks. The text captures the atmosphere of suprise and the subjective impact provoked by those new images in the early 20th century: Castorp's encounter takes place in 1907, at a time when X-rays were still a novelty. One of the main questions that emerges in that chapter concerns the relationship between vision and legibility, and questions the images' alleged neutrality and objectivity. The first time he sees radiographic image, Castorp 'sees' nothing. Only after Dr Behrens, one of the sanatorium's chief physicians, shows him the different anatomical parts he exclaims: 'Yes, yes! I see, I see! . . . My God, I see!' (Mann 1999: 216).

Seeing what is being presented requires a decodifying gaze; a language must accompany the imaging technology and name what is being shown. In a beautiful phenomenological analysis, the episode also depicts the feeling of strangeness, trangression and phantasmagoria that permeate the protagonist's first encounter with the body's interior, as expressed in the singular mixture of his feelings. His 'itch to commit the indiscretion' mingles with 'religious emotion and feelings of concern' (ibid). Contemplating his own hand in the X-ray apparatus confirms the certainty of his own death. In Thomas Mann's (1999: 216) penetrating depiction:

> Hans Castorp saw, precisely what he must have expected, but what it is hardly permitted man to see, and what he had never thought it would be vouchsafed him to see: he looked into his own grave. The process of decay was forestalled by the powers of the light-ray, the flesh in which he walked disintegrated, annihilated, dissolved in vacant mist, and there within it was the finely turned skeleton of his own hand, the seal ring he had inherited from his grandfather hanging loose and black on the joint of his ringfinger . . . With the eyes of his Tienappel ancestress, penetrating, prophetic eyes, he gazed at this familiar part of his own body, and for the first time in his life he understood that he would die.

Mann's commentators have not emphasised this episode.[7] It is nevertheless crucial, since Castorp's perceiving his own sketelon and thereby grasping his mortality have a deep existential and ontological impact on the protagonist's life. From then on, he quit his bourgeois life, cut his links with the world beyond the sanatorium and began the scientific and philosophical studies that would transform him into an intellectual. In a process that recalls the development described in the German *Bildungsroman* or formation novel, his desire to learn about the body ranged from anatomy and physiology to metaphysics. For Castorp, intellectual and spiritual growth were connected to the discovery and knowledge of the body – so much so that his development constituted a displacement, from philosophy to physiology, of the classical *Bildung* ideal. Knowledge of the inner body leads to the knowledge of the self.

What Thomas Mann described was not unusual among early 20th century patients who underwent radiography. 'I feel more knowledgeable about myself. I drew a mental picture of what's in there' (quoted in Kevles 1997: 267): the remark by a man from Chicago confronted with a radiographic image of his body corresponds with Castorp's experience. A century after

Röntgen's discovery, viewing the interior of one's body is still connected to self-knowledge, as expressed by an artist in 1993: 'I had a bone scan last year and was completely amazed at the self-knowledge gained from watching my skeleton emerge on the screen' (ibid). By the late 20th century, imaging technologies had popularised such images. Even those who have never undergone a scan have encountered them in the media. The display of visceral interiority accompanies the externalisation of intimacy embodied in various theatres of the self, from reality shows to the internet publication of personal diaries. Although the displayed images are personal and intimate, in the sense of referring directly to our mortality and corporeality, their propagation and reproducibility turn them into public objects.

A century ago, however, the situation was different. Moral and emotional interiority were shielded from public view; there were clearer boundaries between the public and the private spheres, as well as between the internal and external surfaces of the body. Radiographical images contributed to blur those distinctions and are emblematic of times in which the very notions of privacy and intimacy were beginning to change. Many individuals reacted against the idea that images of their inner body could become public, even if they remained anonymous; as private and intimate, those images had to be jealously protected. Emily Culverhouse's 1897 poem 'Photography' begins: 'An Englishman's body belongs to himself, / but surely that proverb was made / Before Dr Roentgen's impertinent ray'. With wonderful humour, Culverhouse goes on to describe how 'Our dearest possessions are dragged into light', and how, ultimately, the new technology 'will picture the image begot by our brain, / and reveal our inmost perception'. More ominous, but nonetheless revealing of the same phenomenon, is Dr Behrens's remark in *The Magic Mountain*: 'I believe you're afraid to reveal your insides to us, aren't you, Castorp?' (Mann 1999: 212).

X-rays operate a radical change in perception. The medical disciplines were open to the new technology. Although devices to visualise the internal organs were invented in the second half of the 17th century (Segal 1979), major developments would take place only later. The ophthalmoscope was invented in 1850, the laryngoscope in 1857, followed by devices visually to explore the vesicle, the stomach, the rectum or the vagina, which made use of incandescent lamps and photography (Kemp 1998; Reiser 1990). Prolonging the anatomical tradition, these instruments contributed to produce a fragmented and atomised image of the body. Moreover, they privileged vision at the expense of the diagnostic use of the other senses, especially hearing and touch, which had been important in earlier times. Radiography, therefore, did not bring about an epistemological rupture. It did, however, give rise to enormous interest, as attested by the fact that in 1896, a year after the discovery of X-rays, 49 books and over 1000 articles were devoted to them (Gugerli 1999).

Interest in radiography was not limited to the medical world. On the contrary, it spread rapidly into the socio-cultural and juridical fields and penetrated the popular imagination. A few weeks after the discovery, machines were set up in Chicago and Kansas where one could pay to see the bones of one's hand. Patients in Thomas Mann's Berghof carried their own radiographies as if they were wallet photographs; and it is in that form that Castorp kept the 'inner portrait' of her beloved Clavdia Chauchat. Psychologists and parapyschologists welcomed X-rays as a means of exploring the fourth dimension, psychic auras and extrasensory perception. Among spiritualists and students of religious phenomena, some saw in radiographic pictures images of the soul (Bauduin 2012; Cartwright 1997; Grove 1997; Pamboukian 2001). In the midst of a debate on the objectivity of the new images and on the conflict between verbal and visual evidence, X-rays were used for the first time in a court of law in Montreal in 1896, and several cases followed around 1900, mainly concerning medical malpractice. These cases presented a complicated situation insofar as physicians were the accused, while other physicians were called as experts to assess images that remained inaccessible to untrained persons (Golan 1998, 2004; Gugerli 1999; Kevles 1997; Reiser 1990).

Radiographic pictures circulated widely, from specialised knowledge to popular fantasy, from scientific research to various forms of popularisation. They immediately entered literary fiction. In H G Wells's 1897 *The Invisible Man*, the protagonist discovers a ray akin to 'Roentgen rays', which makes his body invisible; in W S Harris's 1905 *Life in a Thousand Worlds*, scientists 'follow the course of a thought in a living cerebrum after the brain has been made visible by a light more potent than the X-ray' (quoted in Kevles 1997: 118). In the visual arts, the ideal of transparency that they seem to convey has a profound impact on the transparent architecture of the German *Bauhaus*, as well as on the Soviet and Italian avant-garde. Bodies are no longer opaque, writes Umberto Boccioni in his *Futurist Manifesto* of 1910; enhanced by technology, the senses have become able to intuit 'the obscure manifestations of paranormal phenomena'; and the artist should therefore create taking into account the new power of vision, 'which may lead to results analogous to those of the X rays' (Boccioni et al 1910).

Rhetoric of transparency

Radiography remained the main device for the production of images of the inner body until at least the end of the Second World War. Starting in the 1950s, its pre-eminence was undermined by other technologies, such as ultrasonography, computerised tomography (CT), magnetic resonance imaging (MRI) and positron-emission tomography (PET). Their reception, however, was not comparable with the speed with which X-rays disseminated or the immediate enthusiasm they unleashed. The first radiographic images were made almost immediately after the discovery of the rays in 1895; Röntgen's results were published in a matter of days; and it took little time for the first applications to develop. In contrast, three decades went by between the first MRI measures in 1946 and the first MRI images of the human body (Dolby and Alker 1997; Gugerli 1999; Kevles 1997).

As we noted, what radiographic images showed was not immediately recognisable. In spite of giving an impression of direct legibility, they required an expert gaze capable of interpreting them. The same applies to later visualisation technologies. They remain surrounded by a discourse about their immediacy, by the rhetoric of self-evidence (Borck 2001) that accompanied the rise of mechanical objectivity in the mid-19th century (Daston and Galison 1992, 2007). The marketing department of the company that developed CT scanners was surprised at the scant interest the new devices inspired in radiologists. But such indifference was not really surprising. As one of the creators of PET, Michael Phelps, noted: 'Anytime that you look at something different [from what] you've ever seen before, you've got to learn what it looks like. And you have to learn what the norm looks like, before you can say, "This is normal"' (quoted in Dumit 2004: 118). And 'normal', as George Canguilhem (1989, 1991) emphasised, is a problematic term: it may designate at the same time a fact and a value attributed to that fact, the result of statistical calculations or that which corresponds to an ideal norm. The history of the visualisation of the body attests to such ambiguity.

The first CT images were an enigma for physicians as much as for radiologists and surgeons. Robert Ledley, inventor of one of the earliest CT scanners, explained that he had to publish his own atlas to teach radiologists to 'see' the images produced by the new technology (Kevles 1997). The difficulty posed by the need for expert judgment is compounded by the fact that, owing to methodological reasons as well as technological and analytical developments, the results of different studies are often hardly comparable and therefore difficult to generalise (Dumit 2004).

Although computerised tomography (CT) operates with the same type of radiation as X-ray technology, the images it generates are of a different nature. CT scanners send X-rays throughout the body via a series of detectors that transmit signals to a computer. A program transforms the signals into pixels and creates a three-dimensional image of the body. The image can be manipulated in various ways as regards colour, size and detail. Signals are different in magnetic resonance imaging, but both technologies share the problem of constructing a three-dimensional image on the basis of information that originates inside the body (Blume 1992; Kevles 1997; Porter 1999; Wolbarst 1999). The production and interpretation of PET images demand a greater ability because while radiological images convey morphology, PET represents physiological function.

The brain regions shown on a PET scan to absorb the radioactive molecules differentially, do not correspond with anatomical regions. Neurologist John Mazziota emphasises: 'A basic premise that must be discarded is that structural and functional anatomy are equivalent' (quoted in Dumit 2004: 196). The difficultes of transforming neural activity in images is a crucial reason for not seeing neuroimages as photographs of brain activity (Roskies 2008). Moreover, anthropologists have noted that the reading of such images usually excludes the social and cultural contexts involved in their production (Cohn 2008). Considering their intuitive visual appeal, their apparently unmediated and familiar character, the images produced by PET and other recent imaging technologies should be approached with caution. Robert Crease (1993: 554) warns against the illusion of transparency and familiarity, against the impression of being before 'a crystal clear perception of what's "really" going on inside the body'.

Medical images seem transparent, inmediate and objective; this impression conceals numerous factors ranging from aesthetic choices regarding the use of colours to make visible neural activity in PET scans to the need for expert knowledges and 'skilful vision' to interpret them (de Rijcke and Beaulieu 2007; Dumit 2004; Joyce 2008; Roepstorff 2007). As a result the epistemological, cultural and technological conditions of possibility of the images are concealed. In a post-photographic era in which all sorts of image-transforming resources are available to any computer user, the production of scientifically valid images must be guaranteed by strict procedures of professional training and control (Dumit 2004; Gugerli 1999). Moreover, to the extent that digital image production technologies are in themselves image-manipulation procedures, they blur the distinction between the original and the copy, since the former already is a technically manipulated image (Borck 2001: 391).

As Barbara Stafford (1996: 133) has observed, reconstructions of the body and the brain obtained through imaging technologies lack 'internal life'; they are 'bereft of recognizable evidence of anguish and pain, either remove the mind from its material circumstances or make the body immaterial'. The resulting images refer to a body that is fragmented, objectified, dematerialised and cut off from its environment – to the body of the anatomo-physiological tradition, deprived of both opacity and subjectivity. It is the body we have, not the body we are. Dumit (2004) uses the expression 'objective self-fashioning' to refer to the assimilation of expert knowledge into the category of the person individuals adopt for themselves. Objective self-fashioning is a double process: on the one hand, science produces facts about what we are; on the other hand, individuals incorporate into their self-perception models of the self based on scientific information transmitted by the mass media. Thus, imaging technologies have become a powerful source for the formation of an objectified self and an objectified body. This is the kind of body of such postmodern anatomies as Bodyworlds and the Visible Human Project, which are discussed in the next section.

Postmodern anatomies

By 2013, more than 36 million people worldwide had visited Bodyworlds (http://www.bodyworlds.com/en.html).[8] The most successful exhibition of all times is based on a taboo: the exhibition of a corpse. Plastination, the technical name of the corpse conservation procedure developed by Gunther von Hagens, has possibly transformed this professor at the University of Heidelberg into the most famous and polemical anatomist since Vesalius and Harvey – ancestors he claims as his own. Von Hagens affirms his right to show corpses, thus returning the public to the performances that predate the confinement of anatomy to students and specialists. He speaks of 'anatomy art', thus linking his project to the particular mixture of science and art, entertainment and medical instruction which characterised anatomy in the Renaissance. Bodyworlds openly recapitulates the history of the anatomical tradition and links it to contemporary somatic culture, thus opening up new possibilities of self-knowledge. In such a context, the interior of the body can compete with, and in some cases replace, the interior psychological world as source of truth and self-knowledge. Bodyworlds thus furnishes a sort of somatic interiority that becomes a source of self-understanding; many testimonies confirm that this is one of its main sources of fascination.[9]

The possibilities to visualise the interior of the body afforded by the new imaging technologies as well as by the postmodern anatomy of Bodyworlds offer new ways of fashioning personal identity. The quest for knowledge about oneself and others in 'the heart' and inside the body loses its traditional metaphoric sense. Viscerality is no more than viscerality; even when narcissistically invested, it is no longer a metaphor for moral or psychological interiority. As already mentioned, Hannah Arendt thought that replacing the richness and variety of public behaviour by the ugliness and sameness of the bodily interior would entail a great loss, not only epistemically and experientially, but also politically. If inside we are all alike, the obsession with visualising the body is a form of body normalisation. And yet, through its artful display of corpses, Bodyworlds counters the depiction of the inner body as characterised by 'monotonous sameness and pervasive ugliness' (Arendt 1979: 35).

While offering itself as a set of artworks, Bodyworlds also conveys the idea of a transparent scientific truth, whereby plastinated flesh offers a direct look 'inside the real body' (Von Hagens, quoted in van Dijck 2005: 56). According to von Hagens, Bodyworlds satisfies the demand for immediate reality and 'unadulterated authenticity' that characterises contemporary culture (von Hagens 2008: 33). The exhibition thus embraces the realistic-moral dimension of traditional anatomy. At the same time, as van Dijck (2001, 2005) notes, it constitutes a meta-commentary on the body and the influence of medical technologies. One of the exhibited corpses, labelled 'orthopedical plastinate', is covered with internal and external prosthesis, thus representing the body as it has become for many in techno-medicalised societies: a cyborg, a mixture of organic and technological parts, neither natural nor artificial – a distinction that, as Bodyworlds proclaims, no longer makes sense (Broadhurst and Machon 2009; Clark 2003; Cleland 2010; Smith and Morra 2006; van Dijck 2005; Zylinska, 2002).

Bodyworlds nonetheless revives a certain type of normativity that can be found in the great Renaissance anatomies. Vesalius's *Fabrica* (Figs. 2.4 and 2.5), Juan Valverde's *Historia de la composición del cuerpo humano* (Fig. 2.6) and Charles Estienne's *De dissectione partium* (Fig. 2.7) use canonical poses from recognised artworks and artistic conventions. As Vesalius explained: 'It is desirable that the body employed for public dissection be as normal as posible according to its sex and of medium age, so you may compare other bodies to it, as if to the statue of Policletus' (quoted in Harcourt 1987: 42). The average normal body corresponded to the anatomical Canon of Policletus. Hence, the illustrations of the *Fabrica* try to remove all traces of individuality and to offer general depictions of human anatomy. The body that is publicly dissected is not a desecrated individual corpse, but the embodiment of a norm to be demonstrated.

In accordance with such tradition, but also with postmodern eclecticism, Bodyworlds incorporates different artistic styles and conventions to produce a pastiche. Some plastinated bodies evoke Vesalius's *écorchés* or Rodin's Thinker. Von Hagens, however, goes beyond simply reproducing past artworks or established conventions. His plastinated bodies are not representations of bodies, but imitations of representations; at the same time, they are bodies made of bodily material; not bronze, marble or paper. The object and its representation fuse in the 'real' plastinated body, and the distinction between original and copy collapses. Von Hagen's plastinated corpses illustrate the fate of bodies in contemporary biomedicine and popular culture, in which the body and its representation, the natural and the artificial no longer operate as orientating references. In this, Bodyworlds expresses the ideal of a transparent, immediately accessible and objectified body akin to that at work in CT, MRI and PET technologies.

The same can be observed of the Visible Human Project. The project is aimed at creating complete, anatomically detailed, three-dimensional representations of the normal male and female human bodies on the basis of CT and MR images of frozen male and female corpses sectioned at millimetre intervals and then further CT scanned and imaged using MRI. Data packages are made available for download after payment.[10] The corpses transformed into virtual bodies can now be unlimitedly manipulated, assembled and disjoined. They constitute a whole new topography to be explored and discovered.

The Visible Human Project (http://www.nlm.nih.gov/research/visible) shares with Bodyworlds the ambition to give back to anatomy the spectacular dimension it had in earlier centuries. The former, however, is based on the corpses of two particular persons, identified as such. The plastinated corpses

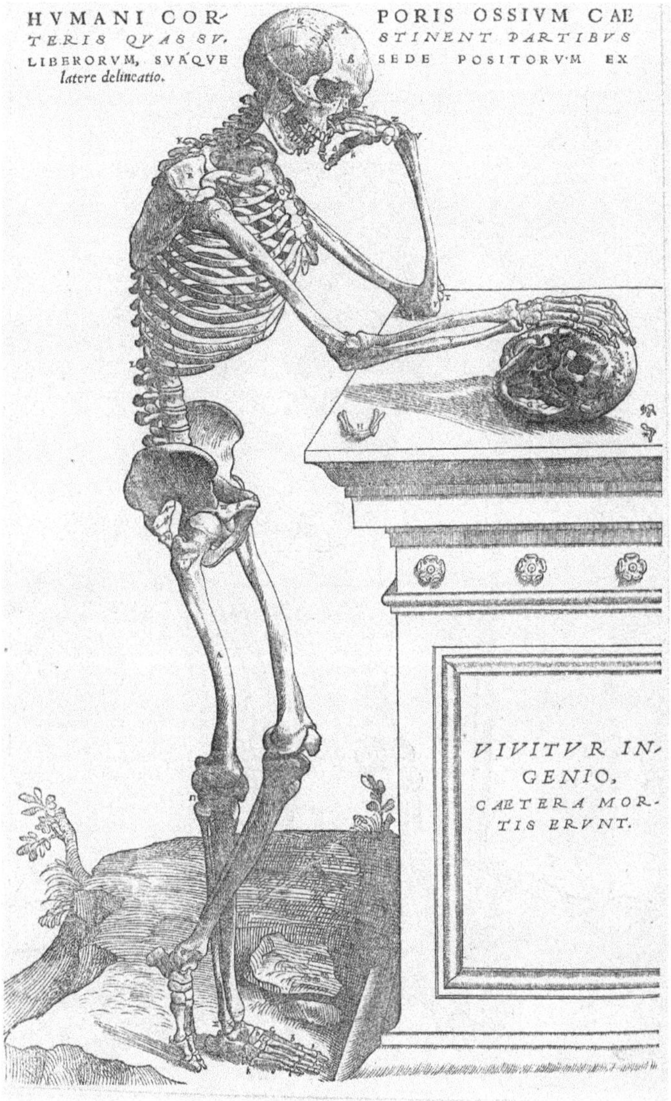

Figure 2.4 Andreas Vesalius, *De humani corporis fabrica libri septem*, 1543
Credit: Wellcome Library, London

Figure 2.5 Andreas Vesalius, *De humani corporis fabrica libri septem*, 1543
Credit: Wellcome Library, London

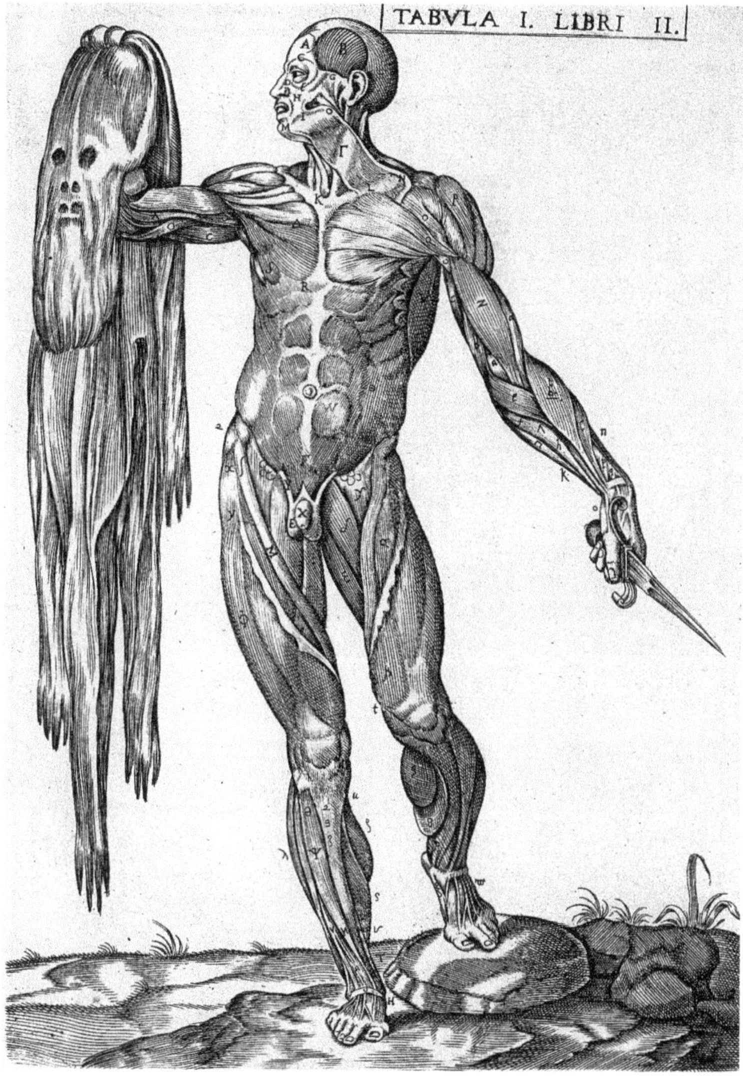

Figure 2.6 Juan Valverde de Amusco, *Historia de la composición del cuerpo humano*: flayed skin, 1568

Credit: Wellcome Library, London

Figure 2.7 Charles Estienne, *De dissectione partium* ..., 1545, Plate showing a reclining male figure with back of head exposed

Credit: Wellcome Library, London

in Bodyworlds are anonymous, yet unique displays, based on individual, irreproducible bodies. In contrast, the bodies of the Visible Human Project are 'digital clones' of once real existing human beings, since the Project has succeded in copying basic features of real material bodies into the digital domain (Waldby 1997, 2000). According to the Project's director Michael Ackerman, virtual anatomy provides a standard model for anatomy – and for medical and scientific training – in the 21st century. The available database enables realistic simulations that are perfectly consistent with the increasingly digitalised practice of medicine and medical education. Diagnostics and treatment have become dependent on imaging technologies, and that determines the specialist's view of the body.

In the Center for Human Simulation of the University of Colorado (see http://www.ucdenver.edu/academics/colleges/medicalschool/centers/Human Simulation/Pages/Home.aspx) the digital corpse has become indispensable. The virtual body, complete with blood circulation and physiological reflexes, helps train physicians in virtual surgeries. Anyone connected to the internet can have access at any time to surgical interventions (Nettleton 2004). Virtual surgeries and telesurgeries make it possible to operate on patients at a distance with the assistance of a robot (Belling 1998; Doyle 2007; Prentice 2005; Thacker 2000; Williams 1997; Williams and Bandelow 1998; Satava 1995). For Richard M. Satava, professor of surgery at the University of Washington:

> . . . the day may come when it would not be possible to determine if an operation were being performed on a real or computer generated patient . . . the threshold has been crossed; and a new world is forming, half real and half virtual.
>
> (Quoted in Csordas 2000: 188)

Surgical simulations and telesurgery highlight the paradox latent in the coexistence of progressive distancing and increased intimacy. Both are far away from the physical body, yet both share the proximity to its interior that technology enables (Doyle 2007; Nettleton 2004). Telesurgery broadcast over the internet is part of the 'cybercasting events' that have become part of daily life (Featherstone 2000), and reinforces the spectacularisation of medicine. There are obviously theoretical and practical differences between the use of imaging technologies in medical practice, the spectacular and artful display of cadavers in Bodyworlds, the surgical interventions available on the internet and the virtual anatomy of the Visible Human Project. The camera can be used for medical purposes or to pursue purely media goals. The critiques addressed to von Hagens and the spectacularisation and commodification of anatomy do not apply to the conventional use of medical imaging technologies for clinical diagnosis. With regard to the display of bodies, the two universes nevertheless share an ideal of transparency and malleability.

From an epistemic point of view, the same kind of body – transparent, fragmented, objectified, commodifed and disembodied – underlies the different technologies and cultural manifestations analysed in this chapter.

The Visible Human Project generates photorealistic images that promise to end the need for interpretation and mediation traditionally required for dealing with visualisations of the body (Waldby 2000). Given the pre-eminently pedagogical function of medical illustration, from anatomical atlases to virtual anatomies, 'legible' images require intepretations, on the part of the researcher or the illustrator, that determine visual emphases or omissions. The result thus involves a balance between reproduction and construction, between an ideal of objectivity and interventions aimed at facilitating the decoding of the images.

Since the Renaissance, the commitment to a naturalistic and realistic depiction of the human body was inexorably related to intellectual, social, visual and artistic conventions. As we have seen, Vesalius provided a familiar, identifiable context for his anatomical illustrations. Even in the framework of mechanical objectivity, various visual devices, such as captions, explanatory diagrams, retouches or highlights on chosen bodily parts, play an important role in the social and intellectual rhetorics of imaging (Kemp 1998). An interpretation-free gaze is a fiction that obliterates the adaptation of the image-production process to visual conventions (Borck 2001). The Visible Human Project's ultimate aspiration is to offer a realistic and unmediated simulation of the physical body – an equivalent of the body that would mingle the real and the virtual, and extend the former in the realm of the latter.

The virtual corpse is an icon of the 'society of the spectacle' (Debord 1999) in which the image competes with the material body to be recognised as real. In such a context, the perception, experience and handling of the body become exclusively visual, and the body itself turns into a 'total optical system' (Waldby 2000: 5). To the extent that visualisation technologies shape both the observer's gaze and the observed object, for many individuals the apprehension of the body is mediated by the television or the computer screen. Such a process parallels the visual objectification operated within biomedicine. Vision is here closest to scientific observation and experimentation; the diminished status of audition and touch, which were crucial in earlier medical practices, implies an increased distance between patients and physicians.

It seems important, not only from the epistemic, but also from the ethical and political points of view, to insist that the body as substrate of existence, as well as subject of sensations, emotions and actions, is not the body of the anatomo-physiological tradition nor of the anatomy spectacles of the new millenium. The acting phenomenological body does not correspond to the infinitely malleable and perfectible transparent body of Bodyworlds and the Visible Human Project. Both projects illustrate the commodification of the body characteristic of advanced liberal societies. Unlike the organ donor, who is represented as altruisitc, the plastination donor appears as a consumer making choices about her or his death. As for the Visible Human Body, it manifests the commodity logic of standardised, detachable and interchangeable organs (Stern 2006).

To conclude this chapter, I will develop some general considerations on three interrelated themes that run through the history of the medical visualisation of the body: the passion for the real, the primacy of vision and the pathos of fragmentation.

Passion for the real in the culture of the spectacle

In the preface to the second edition of *The Essence of Christianity*, the German philosopher Ludwig Feuerbach observed that his age preferred 'the sign to the thing signified, the copy to the original, representation to reality, the appearance to the essence' (quoted in Debord 1999: 11). Such observation, the epigraph of Guy Debord's 1967 *The Society of the Spectacle*, touches on the question of the real and the image in contemporary culture and on their relationships to visualisation technologies. In his book, more topical today than when it appeared on the eve of *Mai 68*, Debord denounced the 'spectacular society' in which individuals substitute with images that which is lacking in their real lives, and thus give these lives unity and meaning. In the process, reality becomes an image, and images become reality. However, insofar as images are an abstraction of reality, the spectacular society becomes alienated from the world. Since Debord's book, such process has intensified.

The proliferation of reality-shows and other forms of video transmission in real time (such as RealVideo surgeries), raise the issue of what Alain Badiou (2004) called the 'passion for the real' that, in an apparently paradoxical manner, characterises a world of virtual realities. Indeed, the society of the spectacle does not wish to give up the real. But we must wonder what the real here stands for, given that the distinctions between the object and its representation, the original and the copy have become obsolete. The postmodern anatomies of Bodyworlds and the Visible Human Project offer the image of the body as if it were the body itself. The virtual body gains in ontological consistency; it becomes material, an image-thing, a body-image. Here, however, the passion for the real is a passion for the real turned virtual. But neither the virtual nor the hyperreal are the real; and neither is the image the thing itself, nor the body reducible to its representation.

Slavoj Žižek has convincingly denounced the false passion for the real as a trick for excluding it. He unmasks the passion for the real as a passion for the image in which the 'ruthless pursuit of the Real behind appearances was *the ultimate stratagem to avoid confronting the Real*' (Žižek 2002: 24). In the fictionalisation of the real in 'reality shows', the protagonists are to be fictionalised and, at the same time, played by the fictionalised individuals themselves. They are fictional characters whose role is to be themselves. Thus, what turns out to be false in reality shows is precisely that which they promise to deliver: reality. Referring to the third series of Big Brother Brazil, journalist Eugênio Gucci and psychoanalyst Maria Rita Kehl (Gucci and Kehl 2004: 171) observe that reality shows

> . . . are the most efficient form of illusion ever produced by mass culture: they sell spectators the faithful reflection of their mediocre lives under the severe protection of 'the laws of the marketplace'. They sell the image of the jungle in which competition transforms human relations – only they raise it to the rank of spectacle.

Such considerations are relevant for analysing the effects of visualisation technologies on corporeality. In the same way that, in reality TV, the real itself turns out be false, in medical images, the body itself is delusory. In imaging technologies, the images of the body are presented as the body itself, as an object that corresponds to that of the anatomo-physiological tradition. From a phenomenological point of view, however, such technologies contribute to a disembodiment of subjectivity and a parallel virtualisation of corporeality. The knowledge about the body provided by imaging technologies contrasts with the body as foundation of our subjective being-in-the-world. And this has important consequences to the extent that, as Nikolas Rose (1998: 69) has written, medicine

> . . . appears to offer us a rational, secular and corporeal solution to the problem of how we should live our lives for the best; of how we might make the best of our life by adjusting it to our truth, by letting medicine enlighten our decisions as to how to live it.

The primacy of vision

The real of the contemporary 'passion for the real' is visible. In the society of the spectacle, the body becomes actual only when it is visualised, mediated by a screen or a monitor. Visibility here depends on the appareance of the bodily image in the field of the other, imaginarily represented by television. Visibility, however, is restricted to those individuals who manage to appear on screen and offer their bodies as ideal models for all. In contrast to political visibility, which depends on action in the public space (Assy 2007), the production of bodies is the production of empty visibility – 'empty' insofar as the visualised bodies are deprived of their materiality.

A main feature of the 'pictorial turn' (Mitchell, 1994) in contemporary culture and the society of the spectacle is that visuality becomes 'the ultimate authority of truth', as Hans Blumemberg put it in connection with modern science (quoted in Gugerli 1999: 132). As we have seen, in anatomy and medicine from Vesalius onwards, there has been a strong link between visuality and scientific truth. Each advance of visuality to the detriment of touch was a step towards the replacement of the patient's physical examination and the direct contact between patient and physician by indirect forms of communication, mediated by increasingly sophisticated technologies.

The fundamental problem here is that vision disembodies. The philosophical tradition considers it the most noble of the senses. Knowledge and mental activities are expressed by means of visual metaphors. That is why traditionally the physician was considered a thinker rather than someone endowed with manual abilities (Porter 1993). The greater dignity of medicine compared with surgery corresponds to the greater philosophical nobility of vision compared with touch. Philosophical activity too is expressed in visual metaphors (Jonas 2001; Le Breton 2003; Leder 1990). Since Plato, philosophy has spoken of the 'eye of the soul' and the 'light of reason' (Blumenberg 1993; Jonas 2001). Vision is the gnoseological sense *par excellence*, insofar as it provides knowledge of the stable properties of things. Visual experience abstracts from reality, and moves things away. Hearing, in contrast, reveals particular events, and touch requires direct contact with the object. Moreover, vision gives the impression of disconnecting sensation and movement, whereas touch is linked to physical movement and the quality of the action: the sensation is different depending on the pressure exerted. Touch thus allows us to experience the materiality of objects as well as causal processes. These disappear from visual experience, which produces a neutral image of the object, unrelated to any physical interaction with it. 'Unlike the probing finger', observes Drew Leder (1990: 118), 'the eyes leave its object of exploration unchanged'.

Corporeal materiality constitutes our inexorable physical location in time and space, in history and culture. On the one hand, it shapes and limits us; on the other, it is the condition of possibility of freedom, of the phenomenological 'I can'. As foundation of our being-in-the-world, the body is irreducible to visual experience. Vision does away with the reference to space and time, and gives rise to a loss of corporeal substance. Hence Hans Jonas's reaction to its primacy when he writes: 'The evidence of sight does not falsify reality when supplemented by that of the underlying strata of experience, notably of motility and touch: when arrogantly rejecting it sight becomes barren of truth' (Jonas 2001: 149).

The fragmented body

The visually apprehended body appears fragmented, rather than as an organic totality. The fragmentation of the body has retained the attention of numerous scholars in cultural history, literary theory and art history among other fields.[11] I see no evidence of the taboo about the topic perceived by Krüger-Fürhoff (2001); on the contrary, it may well be that interest in the topic is related to the belief that the fragment constitutes a form of resistance vis-à-vis the imposition of ideal bodily norms. This is a matter of important ethical consequences.

A recurrent theme in discussions about bodily fragmentation is that of the mutilated body, frequently identified with the grotesque body Mikhail Bakhtin (1993) examined in his famous book on François Rabelais. Such body constitutes an anti-model, a form of resistance to the intact and self-contained corporeality embodied in the classical body which became the ideal in the early 19th century. According to Bakhtin, the adoption of such ideal closed off other bodies and the world. The grotesque body of medieval and baroque popular cultures was permeable to them; its orifices allowed for intense exchanges with the environment and the cosmos. It was an incomplete body, without clear boundaries between the inside and the outside, and between itself and other bodies; its organs manifested a certain autonomy, dissociated from corporal totality.

Towards the end of the 18th century, the grotesque body was replaced by a model of the healthy body – a self-contained unity with boundaries clearly defined by the skin. Thus was born the bourgeois model of the individualised body, the 'total body' erected in medicine as norm of the natural body, and in art as norm of corporeal beauty (Benthien and Wulf 2001; Krüger-Fürhoff 2001; Petersen 2007). The 'total body' or the 'beautiful whole' as normative ideals manifest a quest for harmony, a reaction against the post-revolutionary threats to traditional orders. In such context, the mutilated and fragmented body personified a danger to be combated.

At the same time, in literature, the fragment, which had been used by the French and English moralists, who in turn referred to Pascal and Montaigne, became the romantic genre *par excellence*. In contrast to the finished quality of classical art, the modern work of art emerged as inherently fragmented and incomplete. Against the illusion of a harmonising totality, the Romantics accepted the accidental and involuntary aspects of fragmentation (Lacoue-Labarthe and Nancy 1988; Szondi 1986; de Man 1997). From the German Romantics to Nietzsche and up to Benjamin, Derrida and Blanchot, the fragment remained the favourite means of highlighting the discontinuous, torn nature of reality and the work of art, and of opposing them to the totalising aspirations of Western metaphysics.

During the deep crisis that Western humanism underwent between the 1870s and the Second World War, 'fragmentation' became a recurrent way of describing the collapse of traditional references and the generalised sense of uncertainty. The artist was henceforth capable of capturing only bits and pieces. At the same time, the disfigured, fragmented body emerged as an antidote to the supposed illusions of humanism (Harvey 1990; Moraes 2002; Tronzo 2009). The fragment came to occupy a central place in popular culture as well as in the theoretical forms of thought usually described as postmodern and poststructuralist, where it served to oppose all forms of totality, including corporeal ones. The whole was presented as an imaginary construction aimed at masking the transgressive nature of the fragmented body, which should save us from the body as a universal organic totality. Hence the claim by French philosopher Jean-Luc Nacy, that:

> . . . the parts of the corpus do not combine into a whole, are not means to it or ends to it. Each part can suddenly take over the whole, can spread out over it, can become it, the whole – that never takes place. There is no whole, no totality of the body – but its absolute separation and sharing (*partage*).
>
> (Quoted in Hillman and Mazzio 1997: xxiv)

But in what sense does the fragmented body liberate? Is it really the case that the models of the unified body and organic totality have normative aspirations? On the contrary, has not the fragmented body emerged as an ideal? And is it not the case that, insofar as fragmentation and totality are dialectically connected, the fragmented body supports the ideal of the body as totality? (Wenner 2001) In other words, that the image of the unified body emerges in interaction with that of fragmented corporeality?

As sketched above, the fragmented body does provide bodily norms and ideals: it is the body of the anatomical tradition and of imaging technologies. Both focus on fragments, on bodily parts rather than on bodily wholes; magnetic resonance images appear as 'a world without supports – organs floating on their own in a black sea' (Kevles 1997: 199). Such images give rise to the fiction of bodily parts endowed with autonomous life, as if they could persist without their somatic contexts: foetuses without intrauterine environment, isolated brains that lead to overvaluing that organ, and especially the cortex, as source of the self and all its manifestations (Ortega and Vidal 2011; Vidal 2009). The focus on separate organs parallels the movement of medical specialisation (Weisz 2006). In this respect, contemporary biomedicine, especially in connection with imaging technologies, is postmodern, or shares with postmodernism the desire to break away from notions of bodily totality. Like most scientific research, medical technologies obviously require methodological reductionism and a fragmentation of their objects. The problem arises when reductionism ceases to be methodological, and when fragmentation is asserted as a fundamental property of the lived body, rather than as an experience associated with particular phenomena, such as pain. Popular culture and philosophical theories have come to reinforce the pathos of fragmentation, and so have historical investigations focused on particular organs (Jeggle 1980).

The conceptualisation of the body as fragmented also furthers the commercialisation of bodily parts for transplantation. As Stafford (1996: 143) notes: 'The marketability of dismemberings is accompanied by the growth of regulation and the loss of personal control; we store, retrieve, and purchase somatic fragments in the belief that they can regenerate life and thought'. Biotechnological advances have sustained a market for bodily parts

(Parry 2012). Multinational companies seek to appropriate organs, tissues, genes. Foetal and organ transplantation, reproductive technologies and genetic manipulation have increased the commercial value of bodily parts, and its commercialisation has rapidly become a huge business.[12]

The commodification of bodily parts leaves aside subjective bodily experiences; tissues are extracted, used and patented without reference to the individuals that are their source (Sharp 2000). The body as a totality obviously no longer counts; rather, it appears as a potential obstacle to the commercial exploitation of its parts. As Paul Rabinow shows, industry values not only anonymous bodies, but especially organs without bodies. As he writes: 'it is less "the body" than fragmented body matter which has potential value to industry, science, and the individual'. And he concludes: '[b]y reasons of its commitment to fragmentation, there is literally no conception of the person as a whole underlying these particular technological practices' (Rabinow 1996a: 149).

Margaret Shildrick (2010) has recently argued that contemporary biomedical technologies – especially prosthesis and organ transplantations – have radically contested the image of the unified body. She states that, phenomenologically, the incorporation of biotechnological devices into one's own corporeality opens up new possibilities for intercorporeality. This, however, does not necessarily involve perceiving the body as a set of detached or detachable parts. Prostheses and transplanted organs are incorporated into the body *schema* as a way to expand its limits. In Merleau-Ponty's famous example of the blind person's stick, the world of tactile objects does not begin in the epidermis, but at the stick's end (Merleau-Ponty 2002). Wheelchair users report their chairs are 'who we are, it's an extension of ourselves', rather than places to sit or pieces of furniture. Thus, they explain, when people 'kick our chair, move our chair while we're in it, or touch our chair without necessarily touching our body, there's no difference'. The wheelchair becomes the person and, in the end, a sense of self 'has leaked into a machine' (quoted in Campbell, F. 2009: 54, 55). In Sterlac's performances, prostheses modulate the experience of the world in extreme manners. For example, through a combination of the appropriate sensors and web interfaces, one could touch the nipple of the loved person at the other side of the world, thus procuring forms of 'intimacy without proximity' (quoted in Scheer 2002: 86). Human corporeality may integrate prostheses and other organic or non-organic parts. These parts are assimilated into the anatomy of the lived body; phenomenologically, therefore, they do not correspond to a dismembered and fragmented notion of the body.

Moreover, the body appears dismembered and fragmented only when envisaged passively, from a spectator's point of view (Todes 2001), as a corpse or as an image. From the position of an inactive observer, bodily sensations are understood as being not in the body, but as experienced in the relevant body part. If I feel pain in my foot or arm, I feel it as located in those anatomical parts. I am the one who experiences the pain, but I feel it 'in my foot' or 'in my arm'. In contrast, when the body is active, we do not perceive it as a divisible object, but as an organic and coordinated unity. Todes (2001: 109) gives the example of the runner who, as runner:

> . . . senses himself to be in *one* place—not 'part' here, 'part' there. He senses that he has body *members*, *not* body *parts*. To perceive his members as parts, is to dismember them in appearance. For parts, unlike members, appear capable of being taken a-part without losing their identity.

The body as a coordinated unity is the phenomenological body: the physical location from which we speak, act and know. That body is neither an object of control and surveillance nor a discursive construction, but rather the subject of experience and action.

****

I tried in this chapter to use a phenomenological approach to critically address the objectification and disembodiment of the body carried out in the history of anatomical dissection and through visualisation technologies. The phenomenological dimension of the body is ontologically constitutive of human subjectivity. Phenomenology insists on the reversibility of the physical and lived body, object-body and subject-body. The Western anatomic tradition and medical visualisation, with their emphasis on the machine-body, the corpse-body or the virtual body, move in the opposite direction.

The status of the body in contemporary societies and the impact of medical imaging upon it has been a beloved topic among historians and sociologists of science and technology. By taking a different approach, I wanted to show the potential contribution of phenomenology to this issue by proposing productive insights and forms of critically engaging with the construction of the body within medical technologies. The phenomenological body, as subject of experience, emotions and actions does not correspond to the fragmented and disembodied body conveyed within the history of anatomic dissection and the technologies of visualisation, nor to the transparent and malleable body of postmodern anatomies. If medicine always provided ideal models of corporeality, as well as bodily solutions for conducting one's life in secular times, to stress the materiality and primacy of the phenomenological body constitutes an important not only epistemic but also ethical and political task.

Chapter 3

From ascesis to bioascesis

When useless servitude has been alleviated as far as possible, and unnecessary misfortune avoided, there will still remain as a test of a man's fortitude that long series of veritable ills, death, old age and incurable sickness, love unrequited and friendship rejected or betrayed, the mediocrity of a life less vast than our projects and duller than our dreams: in short, all the woes caused by the divine nature of things.
(Marguerite Yourcenar, *Memoirs of Hadrian*)

The ascetic imperative

In his book *The Ascetic Imperative in Culture and Criticism* (1987), Geoffrey Harpham argued that asceticism is a fundamental instrument in cultural formation and transformation.[1] Any narrow identification of asceticism with a set of restrictive behaviours distinctive of delimited historical periods and particular geographical areas misses the general character of the phenomenon. Even if it exists in all cultures, it does not correspond to any one particular sort of religious experience or set of beliefs and practices, concerning for example the superior value of the soul over the body, or the withdrawal from society (Valantasis 2008). Ascetics may challenge their culture, be integrated within it, transcend it, live in tension with it or transform it (Mitchell 1996). Asceticism may therefore be seen as corresponding to a truly transcultural 'imperative', which can nonetheless be grasped only through the specific forms, motives, contexts and behaviours in which ascetic activity takes place. Michel Foucault came close to this notion when he defined 'practices of the self' as 'models that he [the subject] finds in his culture and are proposed, suggested, imposed upon him by his culture, his society, and his social group' (Foucault 1997a: 291). His history of subjectivity can be seen as a genealogy of ascesis. This chapter draws inspiration from such genealogy by seeking to contrast and compare the ascetic practices of antiquity with contemporary bioascesis and to spell out the main elements of the latter. This approach should serve to highlight continuities and discontinuities and thereby to investigate historical specificities: it will often be the case that the same practices serve different goals and processes of subjectification.

Before dealing with ancient ascetical practices it is necessary to make some terminological observations. The notion of ascesis is frequently associated to the denial, mortification and renunciation of the self. Its valence is particularly negative when seen in the perspective of Christian asceticism. In his work Foucault used *ascesis* (frequently as a translation of the Greek askêsis) and *ascetics* (*l'ascétique*) to stress the positive dimension of self-constitution and self-transfiguration. In his description, ascesis involved acquiring something we do not have rather than renouncing something we do have. Ascesis, he writes: 'does not reduce: it equips, it provides' (Foucault 2005: 320). Some scholars use the term asceticism in a sense close to that of Foucault's ascesis. In this chapter I will use *ascesis* to refer to both Greco-Roman and Christian ascesis. I will contrast it with the notion of *bioascesis*, which I will use to refer to the wide array of reflexive resources – from diet and fitness to psychiatric drugs and cosmetic surgery – that individuals use to act upon themselves in corporeal terms. Bioascesis emphasises bodily, hygienic, medical discourses and practices in ways that are consistent with the neoliberal ideology of the free and autonomous entrepreneurial self.

The outline of the chapter is the following. In the first two sections I will make some general observations about ascesis and asceticism in the ancient world, and then I will sketch some general elements of Greco-Roman and Christian ascetic practices following the work of Michel Foucault, Peter Brown and other scholars. The goal is to highlight the ethico-political dimension of ancient ascesis, or 'practices of the self' in Foucault's words. I will then critically discuss the use of Foucault's terminology to examine neoliberal regimes of self-care. Equating ancient ascesis with neoliberal regimes of self care discharges the former of its ethical significance and obliterates the ethical and political implications of contemporary practices of the self. As a way to differentiate them, I will therefore propose the notion of bioascesis. The following section examines bioascesis in the context of somatic culture and the phenomenon known as healthism. The successive two sections approach the bioascetic somaticisation of subjectivity and the psychological implications of taking the body as matrix of personal identity. I will then explore again the contrast between classical ascesis and contemporary bioascesis, and show the extent to which the 'intimate public sphere' (Berlant) of healthism and bioascesis differs from the ethico-political context of classical ascesis. In the last section I consider body modification as a form of 'somaesthetics' (Shusterman) and of resistance to corporal normalisation operated in healthism and bioascesis.

Ascesis and asceticism

Many historians share a positive understanding of asceticism as involving primarily transfiguration rather than mortification (Valantasis 2008). 'Withdrawal and self-denial were not thoroughly negative categories to those who practiced them', write Wimbush and Valantasis (2002: xxviii). For their practitioners in Greco-Roman and early Christian antiquity, ascetic practices represented beauty, freedom and joy (Ware 2002); hence Foucault's emphasis on the distinction between Greco-Roman ascesis and Christian asceticism. While the latter aimed at detachment from the world, the former comprised an active involvement in it (Foucault 2001: 143–44; see also Heyes 2007; Ramos 1994; Yates 2010). Historians of Christian asceticism contest the opposition and highlight the world-affirming dimension of both traditions; the materials discussed in this chapter tend to confirm their position.

The ascetic phenomenon has four main features, as follows.

First, it implies a process of subjectification, a shift from one kind of subjectivity to another, which is generated precisely through ascetic activity. The ascetic oscillates between an identity to be rejected and another to be achieved. As Valantasis observes, the ascetic 'participates fully neither in one subjectivity (the one left behind, but still being overcome) nor in the other subjectivity (the not-yet-present, but the one on the horizon)' (2008: 43). The ascetic's true identity resides in the desired subjectivity towards which all ascetical practices are directed, and which varies according to historical context. As a result, the same practices may be linked to divergent subjectification processes, such as the constitution of oneself as a moral subject in Greco-Roman antiquity, self-renunciation and purity in late antique Christianity,[2] or bioidentities since the late 20th century. The subjectivities to be achieved through ascetical practices may or may not differ from socially, culturally or politically prescribed identities. While in the Greco-Roman antiquity and early Christian contexts, the self to be attained frequently challenged prescribed identities and embodied forms of cultural resistance; the pursuit of the perfect health and body so often found in contemporary bioascesis usually conveys conformism. This, as we shall see, is particularly so in connection with the phenomenon known as healthism, which has a predominantly self-centred and narcissistic character.

Secondly, ascesis customarily implies the delimitation and restructuration of social relationships, the development of an alternative set of social bonds and the constitution of a divergent symbolic universe. According to Richard Valantasis (2008), co-director of the Institute for Contemplative Living in Santa Fe, New Mexico, ascetic practices are always in conflict with dominant social structures, and one should speak of asceticism only if there is a conscious and deliberate construction of an alternative world defying dominant society. Ascetic conduct involves the intention to 'resist, to contradict, to defy, or to define the self in opposition to the dominant social context' (ibid: 106). Valantasis therefore distinguishes between asceticism and formation. While the latter reconstructs the subject for the dominant society, the former constructs it for an alternative one. Asceticism thus involves difference and dissent. In contrast, formation, also called enculturation, produces conformity and erases difference. Hence, 'only performances intending to inaugurate a new, resistant subjectivity' should properly be 'classified as ascetical' (ibid: 104).

Not all historians of asceticism share Valantasis's interpretation. Keith and Vaage (1999), for instance, stress that in Greco-Roman antiquity, asceticism was not necessarily subversive and could, in fact, help maintain the social order and contribute to the self-assertion of a social, cultural and political elite. The debate continues, and need not be resolved here. For the aims of this chapter, I subscribe to the broader view that does not restrict the ascetic phenomenon to the construction of alternative worlds. I do, however, contrast the explicitly socio-political character of classical ascetic conduct with the apolitical inclination of contemporary bioascesis. Likewise, Foucault (1997e: 271) explicitly contrasts ancient ascesis with the 'Californian cult of the self', in which one is impelled to unravel the true self through psychological or psychoanalytical techniques.

This takes us to the third feature of ascesis, namely that it is not only an individual but also a social and political practice. According to Foucault, among the Stoics 'the care of the self – or the attention one devotes to the care that others should take of themselves' implied 'an intensification of social relations' (1986: 53). Even forms of radical *anachoresis* (withdrawal) presupposed an audience. Furthermore, ascetics played a fundamental role in the constitution of early Christian society. Although frequently described as isolated from the world, most depended on community support and fulfilled important political functions as mediators, conciliators or intercessors at a time when legal and governmental structures were in profound transition. Finally, ascetics stressed group solidarity and shared the values and needs of their community (Brown 1978, 1982; Clark 1999; Rapp 1999; Rousseau 1999).

The fourth main feature of ascesis, in both its classical and Christian varieties, is that it involved the will. The Cynic-Stoic tradition of ascesis as a regimen and exercise of the will (Capelle 1967; Lohse 1969) was incorporated into Christianity as a way of opposing Gnosticism, which was rather linked to the Orphic-Platonic mystical tradition (Foucault 2005). The opposition between right and false uses of the will commands ascetic practice in late antique Christianity. Through the exercise of the will the ascetic may return to the situation prior to the Original Sin. Ascesis constitutes a bodily and spiritual *imitatio Christi* (Brown 2008; Drijvers 1985; Horn 1998). *Stultitia*, or mental restlessness and irresolution, characterises the individual who neither cares for himself nor displays constancy of the will. As Seneca states at the beginning of Letter 52 to Lucilius, while the *stultus* has a weak, fragmented, limited will (*akrasia*) and is inattentive and distracted, unable to persevere in the practice of ascesis, those who possess free will must be capable of attention, surveillance, constancy and athletic concentration (Foucault 2005: 132–35; Voelke 1973).

As noted, Foucault uses the words 'ascesis' and 'ascetics' to avoid the negative connotations of renunciation and mortification involved in the term 'asceticism'. Ascetics are 'the more or less coordinated set of exercises that are available, recommended, and even obligatory, and anyway utilizable by individuals in a moral, philosophical and religious system in order to achieve a definite spiritual objective', namely transforming oneself into a subject of action and true knowledge (Foucault 2005: 416–417). Valantasis's view partly overlaps with Foucault's. Both refer to ascetics in a positive and productive sense, and both stress the element of performance and practice, as well as the process of subjectification they involve. In the end, Greco-Roman and early Christian ascesis was an 'exercise of the self on the self' (Foucault 1997a: 282), but was also endowed with a social and political dimension, and implied action in the public realm.

Ascesis as a practice of freedom

Various Greek and Roman philosophical schools and Christian asceticism share a dual understanding of ascesis. For cynic philosopher Diogenes, body and soul have to be trained simultaneously; ascesis of the one 'was worthless without the other, good health and strength being no less useful than the rest, since what concerns the body concerns the soul as well' (quoted in Foucault 1990: 73). Even corporeal forms of ascesis involved a spiritual dimension. For classical Greek thinking, ascetic practices were integrally linked to the *polis*; the individual's self-attitude contributed to the 'well-being and good order of the city' (Foucault 1990: 79). Dietetics was a fundamental part of Greco-Roman ascesis, encompassing not only diet but also sexual habits and exercises, as a well as a set of other behaviours (Holmes 2010). It continued to prosper in the fourth century and in the Hellenistic and imperial periods analysed by Foucault, constituting an important part of political life.[3] Dietetics involved a return to the natural order consistent with the values and needs of the *polis*. It opposed the excessive cultivation of the body, which would only result in indolent men not useful to the city, and interfere with the care of the soul. The ancient 'valetudinarians' replaced *amor mundi* and a concern with the common good with the narcissistic pursuit of a longer life and maximal corporeal performances, thereby forgetting that to live fearing death is to live fearing life.

As feminist critics of Foucault have emphasised (Foxhall 1998; Richlin 1998), dietetics was a male concern. It was, moreover, an elite affair. Hippocrates highlights that since commitment to health demands total dedication, only a small group of people who are 'well off and convinced that there are no benefits of wealth or anything else without health' can pursue it (quoted in Holmes 2010: 180). Plato notes the wastefulness of prescribing a long regime to a carpenter who lacks the necessary leisure to follow it. As he states in *The Republic*, only the wealthy have time for virtue.

Foucault's interpretation has raised objections. Wolfgang Detel argues, on the one hand, that the valetudinarian excesses Plato criticises for being obstacles to study and to proving one's talents (2005: 407b–c) are not opposed to the dietetics Foucault analysed as an 'art of existence'. On the other hand, he goes as far as arguing that a great deal of ancient dietetics and practices of the self evoked 'a desperate programme of self-help rather than an aesthetic stylisation of life as a whole' (Detel 2005: 96–97). The practical interest in regimen and self-care was, according to him, a result of a lack of doctors and of the terrible reputation of the medical profession. As a consequence, Foucault's description of dietetics as self-stylisation has a relatively restricted application. It concerns only the 'excessive prophylactic diet' which Plato and Aristotle criticised, but which was of 'interest to free and affluent male citizens, and thus pre-supposed those "unequal, hierarchical and authoritarian institutions" that were generally required for scientific activity in the ancient world' (ibid: 97). In its Pythagorean origins, dietetics indeed constituted an art of existence. However, by the time of the classical period Foucault examines, only the decadent variety survived, connected to political privileges and highly asymmetrical power relations. At the same time, Detel argues, dietetics for the masses severed its links with ethics.

Although important, such criticism does not affect our approach here, which consists of bringing out, for the purpose of historical comparison, some general elements of Greco-Roman and early Christian asceticism that related care of oneself with care for others and for the *polis* as a whole.

Like Greek and Roman ascesis, early Christian ascetic practice involved a significant social and political dimension, as well as relational aspects (Rubenson 2002; Valantasis 2008; Ware 2002). Even hermits were not isolated; they had an audience and were surrounded by a community that supported them. Paradoxically, for St Anthony in third-century Egypt, St Basil of Caesarea in fourth-century Cappadocia, or Benedict of Nursina in sixth-century Italy, *anachoresis*, the practice of withdrawing from society, had a world-affirming meaning. These holy men transformed the world around them by way of their solitary self-transfiguration. Historian of religion Kallistos Ware observes that when St Benedict retired to a cave near Subiaco, his quest for salvation had a salutary effect on the community around him. 'Often', he writes, 'it is precisely the men and women of inner stillness who in practice bring about the most far-reaching alterations in the society around them' (Ware 2002: 6). Far from being a selfish quest for individual salvation, ascesis encompassed the whole of humanity.

In the Mediterranean world of the second and third centuries, divine and supernatural forces manifested themselves to a few elected individuals – 'friends of God' and 'holy men' who enjoyed special intimacy with the holy (Brown 1982; Cameron 2012; Hayward and Howard-Johnston 2000). Their affinity to God granted them status and power; they embodied the hope of the community and testified to validate the possibility of approaching the holy through rigorous asceticism. The fourth century 'man of the desert' undertook painstaking work on himself. He had to display a singular lifestyle which the community could easily identify. What makes a performance ascetical is the intention of the practitioner, which has to be subjected to public scrutiny. Valantasis (2008: 39) speaks of the 'performative visibility' of the ascetics, and Peter Brown (1982: 135–36) observes that:

> It is perhaps one of the most faithful indications of the whole style of late Roman society that the objectivity that men so desperately needed was less often vested in impersonal institutions, but was only thought acceptable in a man who could be closely observed to be in the act of forging total dissociation in himself, by hammering it out like cold metalwork, from a lifetime of asceticism.

Ascetics' political role in late antiquity consisted in being considered paradigmatic examples of conduct. At a time when institutions and legal structures did not aim to satisfy individual needs, ascetics fulfilled the function of mediators, examples, intercessors and arbiters who, through their action, contributed to the Christianisation of the late Roman Empire (Drijvers 1984, 1985; Rapp 1999).

Such observations complicate Max Weber's view of early Christian asceticism in *The Economic Ethics of World Religions* 1915–16.[4] In Syria and Egypt priests and bishops were frequently ascetics or exalted ascesis. The Gallic episcopate was sometimes recruited in monastic circles (Brown 2008; Drijvers 1985; Markus 1998). After Constantine's conversion and the end of persecutions, it efficiently replaced martyrdom as a sign of holiness (Conrad 1995; Markus 1998). Moreover, asceticism, specifically sexual renunciation, represented a '*carrière ouverte aux talents*', by means of which uneducated women and men could attain a reputation normally reserved to cultivated male citizens (Brown 2008: 61). As respected and admired public personalities who led an inspiring life of *imitatio Christi*, ascetics did not oppose the institutions of their time, but promoted their transformation. They expressed the Christians' desire to dissociate themselves from pagan religious and social traditions and institutions, and to develop alternative forms of subjectivity and sociality. Early Christian asceticism thus provided a manifest model of the Christian identity and, as Samuel Rubenson (2002: 55) put it, aimed at the 'creation of a *polis* out of the desert'. He goes as far as noting that the 'success of Christianity could hardly have come about without the devotion of the ascetics' (ibid). Averil Cameron (2002) speaks of 'ascetic closure' to describe the predominance of asceticism as a dynamic factor of historical change.

To sum up, in both the Greco-Roman and the early Christian contexts, corporeal ascesis was linked to spiritual ascesis, yet was at the same time profoundly linked to politics and the life of the community. Both ascetic cultures took the body as basis for the constitution of a self apt to participate in public affairs, pursue self-knowledge or annihilate itself in the search for

God. In contrast, as we shall see next, the neoliberal practices of the self that I subsumed under the label of bioascesis do not display the predominant ethical character of ancient asceticism.

Neoliberal ascesis

As in antiquity, ascetic practices in 'advanced liberal' democracies[5] offer a high degree of heterogeneity. Clearly not every form of contemporary ascesis aims *exclusively* at an individual and apolitical self-improvement. Many practices that could be recognised as bioascetic embody commitments to social networks and communities. That is why Valantasis (2008) is able to construct a genealogy of asceticism that begins with ancient philosophical and early Christian ascetics, and from Nietzsche to Jean Genet and beyond, reaches gay activism, feminism and the 9/11 terrorists.

Nevertheless, if Gillian Lindt is right, contrary to ancient asceticism, ascetical practices in contemporary American culture are 'never defined in terms of actions that seriously challenge dominant values of the society or its status quo', and therefore 'play a decidedly limited role in defining the central values of the dominant society of modern Western capitalism' (Lindt 2002: 596). On the one hand, they have undergone a process of increasing privatisation and lost their historical function of contributing to shape a communal and shared cultural identity. On the other hand, ascetical practices related to individuals' health and wellbeing play an important role in the neoliberal context of consumer citizenship.

In his genealogy of the modern subject, Foucault places greater weight on epistemic continuities than on discontinuities. He now finds 'strict continuities' and 'direct borrowing' between early Christian and ancient philosophical ascesis (Foucault 1990: 15).[6] The same practices emerge in different socio-historical contexts, where they give rise to different forms of self-embodiment. Scholars have freely applied the Foucauldian notions of care and technologies of the self to contemporary phenomena and processes of subjectification. These concepts are frequently taken up in governmentality literature, and specifically connected to issues of health and medicine (Race 2009a, 2009b). The result has been an emphasis on the continuities between different socio-cultural contexts, as well as the identification of technologies of the self with neoliberal regimes of self-care as one of the favoured forms of governing individuals in contemporary societies.[7] Canadian philosopher Cressida Heyes argues for example that one key tool of ancient technologies of the self, the *hupomnemata* – personal notebooks for transmitting advice about dieting – have been incorporated in the neoliberal forms of care of the self. She draws four parallels between *hupomnemata* and several Weight-Watchers' strategies, speaks of 'Weight-Watchers *hupomnemata*' and notes that Foucault 'could be running a Weight Watchers' meeting' (Heyes 2007: 85; see also Valverde 2004: 84).

The view of the individual as a self-managing, risk-calculating enterprise mirrors public health policies as well as the economic rationalities of private companies. Governing in advanced liberal democracies demands individual self-care, self-examination and self-improvement (Miller and Rose 2008; Nettleton 1997; Petersen 1997; Rose 1999). Contemporary self-health techniques 'share much with the self-control achieved by body regimen in ancient Greece, and similarly, demonstrate one's ethical worth' (Bunton 1997: 238). In *The Politics of Life Itself*, sociologist Nikolas Rose examines diverse forms of biological citizenship in genetics, reproductive technologies and psychiatry. In his view, although some may dismiss these techniques as 'narcissistic self-absorption', they deploy an 'admirable ethical seriousness'. Moreover, 'like these techniques found among the Greeks . . .', he observes, 'they identify an aspect of the person to be worked on, they problematise it in certain ways, they elaborate a set of techniques for managing it, and they set out certain objectives or forms of life to be aimed for' (Rose 2007: 146).

Nevertheless, as Rose himself acknowledges, activism and responsibility vis-à-vis one's biomedical condition can become a norm, an obligation, rather than a conscious individual choice, as was the case of classical ascetical practices. Thus, regulating diet, pursuing healthy habits and lifestyle to minimise illness and maximise health may turn into a duty. Active and responsible individuals must engage in 'a constant monitoring of health, a constant work of modulation, adjustment and improvement in response to the changing requirements of the practices of his or her mode of everyday life' (ibid: 223). Practices of the self can become practices of discipline and normalisation, as well as ways of stigmatising the individuals who do not follow them.

There is much to be learnt from approaches that draw on Foucault's interpretive vocabularies and establish parallels between ancient and contemporary ascetical discourses and practices. However, Foucault's 'care of the self' should not be equated with neoliberal strategies of self-governance and self-improvement. Nor is it certain that neoliberal governmentality incorporates and develops the language and practices of the care of the self. For example, Cressida Heyes considers that ascetical practices are 'fully recruited again to the service of dieting and attaining normative femininity' (2007: 87). Ascesis is thus coopted by contemporary normalising technologies (ibid: 11). It therefore differs from ancient technologies of the self, which, in Foucault's view, represented an alternative to contemporary forms of governmentality. Several commentators endorse this interpretation:

Heyes (2007: 79) herself states that Foucault returns to Greco-Roman antiquity to discover forms of living that 'although inevitably implicated in disciplinary practices, cultivate a broader repertoire of human possibilities instead of increasing docility'. Mariana Valverde (2004: 82) finds that 'dialogical practices involving peer support . . . constitute a site for askesis, or to put in Foucauldian terms, a venue for developing and using certain technologies of the self'. And Edward McGushin (2007: xx) writes that 'care of the self offers a possible counterpractice in the form of alternative techniques of self-fashioning', or 'disciplines of resistance'. Finally, Foucault's mentor Georges Canguilhem (1997: 32) puts it more straightforwardly: 'In the face of normalization and against it, *Le Souci de Soi*'. Hence, identifying ancient ascesis with neoliberal self-care regimes deprives the former of its ethical significance.

Similar practices in classical ascesis and contemporary bioascesis have opposite ethical goals and result in divergent processes of subjectification. In Greco-Roman and early Christian ascesis, the self to be attained differs from the autonomous and responsible individual of advanced liberal democracies. Foregrounding this distinction may be useful to uncover the ethical implications of contemporary regimes of self-care.[8]

Bioascetic practices and the constitution of bioidentities

As already mentioned, contemporary ascetical practices or bioascesis arise in the socio-political context of advanced liberal democracies.[9] Briefly put, the passage from welfarism, which involved governmental intervention into citizens' wellbeing, to neoliberalism, which renounces such intervention, shifted onto the individual practices of governance no longer managed by state programmes (Miller and Rose 2008). Individuals are then held responsible for optimising their health and quality of life, and active entrepreneurship regulated by market principles replaces passive state solidarity. The discourse of the free and autonomous entrepreneurial self pervades neoliberal political rationalities and state programmes. Individuals must act upon themselves as free and responsible agents (Rose 1999: 68). Governing through the regulated and accountable choices of autonomous individual citizens amounts to conceiving the political subject as 'less a social citizen with powers and obligations deriving from membership of a collective body than an individual whose citizenship is active' (Miller and Rose 2008: 82). This citizenship, note Miller and Rose (ibid), has to be displayed 'not in the receipt of public largesse', but rather 'in the energetic pursuit of personal fulfillment and the incessant calculations that are to enable this to be achieved'. Governing, in turn, takes place 'at a distance' and 'through freedom' (Dean 1994). Discussions about biosociality and biological citizenship have emerged in connection with such political developments.

Paul Rabinow created the concept of biosociality in the early 1990s to link different domains of practice, including genomics, pastoral care, patients' rights and researchers' alliances. His goal was to examine how disease-related socialities change as a result of the new genetics (Rabinow 2008: 190). Biosociality contrasts with older forms of identification among those sharing a disease or biological condition, such as the patients suffering from TB in the luxurious Swiss sanatorium elegantly depicted by Thomas Mann in *The Magic Mountain* (1999). The concept has been widely used in the social sciences to analyse new forms of identity practices and their relation with older modes of classifying people; the role of state and non-state actors and patient organisations; and, finally, the generation of wealth and commercial interests in biology (Gibbon and Novas 2008: 5). In this chapter, I will use the notion of biosociality without reference to genetics, patient advocacy and activism, and will stay closer to those who have employed it to trace the emergence of new identity practices which may interact with older categories such as class, kinship, gender or race. In some cases, these categories apply and may even be reinforced through combinations with new biosocial criteria of identity formation (Gibbon and Novas 2008). For instance, research in the field of pharmacogenomics has led to the manufacture of drugs for specific ethnic and racial groups, as is the case of BiDil, the first pharmaceutical specifically to treat heart failure among Afro-Americans. Supporters of such products argue that the benefits for the target population override concerns about reinforcing race as a biological category (Kahn 2010; Roberts 2010).

Something similar has been noted in connection with gender identities. New knowledge and technologies associated with breast cancer genetics, specifically BRCA susceptibility genes, may end up strengthening gender stereotypes (Gibbon 2008). A related illustration is offered by breastfeeding awareness campaigns, which frequently exploit non-conclusive research on the benefits for the baby and the mother (Wolf 2010, 2011). These few examples, to which many could be added, attest to how new biological identities may interact and strengthen traditional forms of identity constitution. Biological identities may also emerge without directly reinforcing or biologising traditional identity and community forms, or without pursuing explicitly political goals. However, health and fitness magazines frequently reproduce class, gender, sexual and race stereotypes, marginalise race and ethnic minorities and foster body normalisation (Dworkin and Wachs 2009).

Various wellness and holistic practices such as yoga and meditation deploy a spiritual dimension usually connected with Buddhist spirituality (Craig 2002; Hall 2003). These practices may result in an apolitical or not explicitly political sociality, sustained by private individuals and groups that are not organised according to traditional grouping criteria such as race, class, gender, kinship, religion, social function or political orientation. Those individuals and groups assemble around issues of health, bodily performances, specific illnesses or longevity, and they function according to new criteria of merit and recognition, new values based on hygienic rules, time-occupation regimes and new ideal models of the self based on physical features. In such a context, individual actions are conducted with the aim of obtaining better physical shape, living a longer life or prolonging youth (Bordo 1993; Edgley 2006; Edgley and Brisset 1990; Gori and Del Volgo 2009; Hesse-Biber 1997;

Sfez 1995). 'The ideal here is of a body that is absolutely tight, contained, "bolted down", firm, in other words, a body that is protected against eruption from within, whose internal processes are under control' (Bordo 1993: 190).

In this context, a medical-physicalistic vocabulary (about risk and susceptibility, prediction and prevention, involving not only genomes but also cholesterol rates, muscular tonus, physical performance, aerobic capacity and other measures) popularises and enacts quasi-moral norms (Brandt and Rozin 1997), and provides classificatory and evaluation criteria. Individuals use these biological vocabularies to speak and act upon themselves and their relation to others (Rose 2007: 140–41). At the same time, social, religious, sport and sexual activities are reconceptualised and given new significance as health practices. Depending on the situation, psychological notions of personhood may coexist, appear displaced or even in conflict with somatic/cerebral 'bio-identities' or individualities. For example, in a study of the online discussion group *alt.support.depression.medication*, anthropologist Emily Martin describes the clash between a dominant reductionist model of depression, and the participants who challenge the idea that the minds and experiences of people with mood disorders can be reduced to brain dysfunction. Thus, one member of the group, Haily, argues that depression 'isn't a medical illness' but 'a mental illness', and declares '[i]t's not all just science and mathematical equations'. In depression, she writes, '[i]t's not just the chemical reaction, it's what "causes" that chemical reaction, that should be addressed'. Eric, another member, responds that putting 'your emotions into a mathematical equation' is 'very easy' since they 'come from your brain' (Martin 2010: 373–74).

The clash between Haily and Eric illustrates how corporeal forms of subjectification coexist but also collide with intimistic and internalistic modalities of self-construction and description. Diverse forms of subjectification may subsist in the same context, as shown by Martin's ethnography or Callon and Rabeharisoa's (2004) study of two brothers from Reunion Island who suffer from Limb Girdle Muscular Dystrophy. While Leo belongs to a patients' organisation that tries to raise funds, spread information about the disease and gain political recognition, his brother Gino refuses to subjectify himself in terms of his disease. He shows total lack of interest in genetics and the patients' organisation and prefers to spend his time going to a football match. Individuals describe themselves in biological terms, but they also contest and sometimes actively oppose such descriptions. Moreover, since identities are plural and multiple, the same persons speak and act in one context as biological or somatic individuals, while in others they draw on the vocabulary of psychology and intimacy (Rose 2007: 113).

The process of thinking, speaking and acting upon oneself as a corporeal being takes place by means of the wide array of reflexive resources and practices I called bioascesis. To address this phenomenon of personhood being shaped in corporeal terms Nikolas Rose speaks of 'somatic individuals'; e.g. 'beings whose individuality is, in part at least, grounded with our fleshly, corporeal existence, and who experience, articulate, judge and act upon themselves in part in the language of biomedicine' (Rose 2007: 26).

Rose is chiefly concerned with the constitution of genetic identities and neurochemical selves, that is, with the depiction of life at the molecular level, which he contrasts with the molar level. The latter constitutes the body of the 'clinical gaze' depicted by Foucault, and encompasses diet, fitness and cosmetic surgery. Rose's analysis privileges the molecular over the molar. In contrast, I will here focus on the impact of contemporary 'molar' discourses and practices of the self – which I called bioascesis – upon the constitution of 'bioidentities'. I borrow this term from Waldby and her colleagues who, in connection with the circulation of human body tissues, describe bioidentity as the 'commonsense understanding of our bodies as "ours", as both supporting and being included in our social and subjective identities' (Waldby et al 2004: 1462). In this chapter I use 'bioidentity' in a broader and looser sense to describe the process of subjectification involved in diet, fitness, body modification and related self-practices.

Healthism, somatic culture and the new deviants

Within the context of biosociality and somatic culture, subjectification involves acting upon oneself in corporeal terms. Bioascesis, as we saw, reproduces at the subjective level the rules of biosociality and somatic culture, thereby emphasising bodily, hygienic, medical discourses and practices upon self-constitution, and incorporating the ideology of the free and autonomous entrepreneurial self. The government of freedom involves the deployment of 'technologies of responsibilization' (Rose 1999) as a way of controlling individuals without being responsible for them. In this context responsibility becomes a problem of self-care. Responsibilisation is crucial in healthism, which fosters individual responsibility and blame for one's health.

Robert Crawford is credited as the first person to have used the term 'healthism' in 1980 to describe the moralisation of health within the American middle class. This phenomenon represents one important aspect of the medicalisation of everyday life fuelled by health promotion strategies (Crawford 1980, 2006). In 1994 Petr Skrabanek took up the concept in *The Rise of Human Medicine and the Rise of Coercive Healthism*, but he conceived it as a form of totalitarianism which justifies racism and eugenic practices. While Skrabanek linked healthism with state interventions that coerce people into healthy lifestyles, Rose (1999) considers it a technology of responsibilisation without governmental coercion. In short, personal responsibility for health has become 'the sine qua non of individual autonomy and good citizenship' (Crawford 2006: 402).

Since Crawford's first formulation of the topic in the 1980s, healthism has reappeared frequently in medical and public health journals (Greenhalgh and Wessely 2004). It has been related to the 'ideology of conspicuous consumption' known as consumerism (ibid: 197). In the 1990s, discussions of healthism loosely merged with issues of medicalisation, risk and governmentality (Aïach 1998; Bunton 1997; Conrad 1995; Druhle and Clément 1998; Faure 1998; Greco 1993; Morris 2000; Petersen 1997). In more recent years, healthism has been associated with manifold questions: the global context of public health; older adults living with multiple chronic conditions; notions of chronic illnesses; discourses of responsibility in women's magazines; food choice and food behaviours; body ideals in the printed media; and beliefs, behaviour and expectations of health-aware British middle classes (Cheek 2008; Clarke and Bennett 2012; Crawford 2006; Galvin 2002; Greenhalgh and Wessely 2004; Henderson et al 2009; Roy 2008; Rysst 2010). Healthism has been also linked to the impact of economic globalisation, said to perpetuate poverty-related issues among less favoured classes while enhancing the demands for health and body image among the middle and upper classes (McMichael and Beaglehole 2000). Moreover, it has been criticised as a 'new conservatism' prominent in the most developed countries (Cheek 2008: 976). There, the promise of perfect health and body has become a 'badge of honor by which we can claim to be responsible and worthy both as citizens and individuals' (ibid).

For Adele Clarke and her colleagues (2010a), healthism is linked to late 20th century biomedicalisation, which casts health as a moral issue of continuous self-transformation (Clarke et al 2010b: 63). Crawford (1980: 380–81) had already noted the phenomenon in the 1980s when he described the way health became 'a pan-value or standard by which an expanding number of behaviors and social phenomena are judged', and concluded that 'good living is reduced to a health problem, just as health is expanded to include all that is good in itself'. Healthism presupposes and fuels the notion of free and responsible subjects engaged in practices that maintain or enhance good health and body shape (Greco 1993; Harding 1997; Roy 2008). Strength, youth, longevity, performance, health and beauty become normative criteria that measure value and constrain individual behaviour. Our purpose here is to look at healthism as a privileged site where bio-ascetical discourses and practices unfold in advanced liberal democracies to constitute somatic individuals and bioidentities – or, in Alan Petersen's (1997: 198) words, as a privileged form of 'bodyism' that merges hedonistic lifestyles with preoccupation with ascetic practices directed to ideals of perfect body and health.

In such a bioascetical context, illnesses and disabilities that distort the body emerge as signs of personal defeat and failure. The obsession with perfect health and body 'is a secular religion from which the disabled and the disfigured are, of course, rigorously excluded, unless they are willing to play the ossified role assigned to them in reality-based drama as gutsy models for "personal adjustment, striving, and achievement"' (Morris 2000: 159).

A good illustration of this tendency is the figure of Christopher Reeve, the actor of Superman, who, after a horse-riding accident left him tetraplegic, came to embody the 'I-will-walk-again' rhetoric. Disability rights activists noted that Reeve's experience did not match that of many disabled individuals who lead a meaningful life in spite of their impairments, and criticised him for raising false hopes (Goggin and Newell 2009; Molsberry 2009; Peace 2009; Schatzlein 2009; Williams 2009).

'Ableist' attitudes, which devaluate disability vis-à-vis 'normal' able-bodied people, frequently take a paternalist turn. Paternalism, observes political scientist and disability rights scholar Harlan Hahn, allows dominant classes to demonstrate sympathy for minority groups while simultaneously maintaining them in economic and political submission. It thus allows 'the non-disabled to act as the protectors, guides, leaders, role models, and intermediaries for disabled individuals who, like children, are often assumed to be helpless, dependent, asexual, economically unproductive, physically limited, emotionally immature, and acceptable only when they are unobtrusive' (quoted in Campbell, F. 2009: 19). However, inducement to cooperate with surgeries or other treatments to overcome disability do not necessarily result from the 'hegemonic compulsion to ableist normativity', but may follow a sense of responsibility and perceptions about the 'correct' ethical conduct of a 'proper' citizen (Campbell 2005a: 119–20). Nevertheless, the legitimate goal of minimising illness, the search for cures and the pursuit of health can under no circumstance justify denigrating or misrepresenting the life experiences of disabled individuals. Stigma, anger, avoidance and blame towards individuals with disability derive from ideals of bodily perfection, and from a refusal of human imperfection (Martz 2001). As disability scholar Tom Shakespeare argues, to accept limitations and to learn to live with them is more important than the obsession with health and bodily perfection, 'ultimately unattainable or false goals' and 'harmful to well-being and self-esteem' (Shakespeare 2006: 117).

In the bioascetical context of healthism, failure to maintain the ideals of health and bodily perfection manifests akrasia, understood as weakness of will: 'The only tyrant you face is your own inertia and absence of will', writes Donald Ardell, one of the leading figures of the wellness movement, in his best-seller *High Level Wellness: An Alternative to Doctors, Drugs, and Disease*. And he continues: 'your belief that you are too busy to take your own well-being into your own hands and that the pursuit of self-health through a wellness-promotive lifestyle is too hard, complicated, or inconvenient' (quoted in Crawford 1980: 379). The gurus of healthism emphasise our responsibility for our diseases. The *stulti* and the akratics therefore deserve their illnesses, and are a legitimate object of social ostracism and moral condemnation. The basic motto of the English-Canadian women's magazines published between 1997 and 2000 and analysed by Roy (2007) turns out to be: 'Take responsibility or suffer the consequences'. Such a position prolongs the attitudes Edgley and Brisset observed over two decades ago when they wrote: 'Drunks, gluttons, smokers, and sedentaries (now derisively called "couch potatoes" in the new pejorative of healthism)' are considered inferior classes and the uneasiness in the presence of unhealthy individuals frightfully reminds the 'disquiet experienced in former times by good white folk when in the company of blacks' (Edgley and Brissett 1990: 263; see also Crawford 1994).

Healthism as privileged site for the deployment of bioascetical discourses and practices epitomizes the neoliberal ideology of individual responsibilisation, and thinness is its preferred embodiment – both a requisite and an emblem of success, as well as a symbol of merit (Guthman 2009: 193). Fat people are condemned for failing to live up to such standards. Statements such as US Surgeon General Richard Carmona's, comparing obesity rates in America with the September 11 attacks and describing rising weights as the 'terror within' (quoted in Kirkland 2008: ix), illustrate how discourses about an obesity epidemic have become a major technique for the neoliberal governmentality of human bodies. In a 'fat-hating society', writes Marilyn Wann (2009: xiv), 'when fat people do fall ill, we get blame, not compassion. We receive punishment, not help'. In the last decade Fat activists, Fat studies scholars and Fat acceptance groups such as Health at Every Size (HAES)[10] strongly repudiate what they consider a 'witch hunt on fat people' (ibid: x). They believe that the reports of harm caused by obesity are exaggerated. In a frequently cited article in the *International Journal of Epidemiology* Paul Campos and his colleagues (2006) heavily criticise empirical claims about the 'obesity epidemic', as well as the description of obesity as a pathological major cause of mortality, and thinness as good and practical. They problematise statistical linkages between body mass and mortality, which disregard important confounding factors such as fitness, family history or diet quality; they argue that the data available show that even at a BMI of 40 or more, mortality differences are not statistically significant; and they suggest that many health problems of fat people are probably the result of yo-yo dieting and diet drugs. They conclude that the public rhetoric about an obesity epidemic is being driven by cultural and political factors, rather than by any concrete evidence. Commenting on those studies, communication scholar Kathleen Lebesco assumes a 'distinct lack of proportion between the hype surrounding obesity and the lived experience of fatness' (Lebesco 2010: 74).

Paul Campos argues that people who express negative attitudes towards fat people are more likely to express negative attitudes towards black and poor people; their aversion is a new way for the upper classes to show their rejection of the lower ones, and has nothing to do with statistics and everything to do with 'the moral superiority engendered in thin people by the sight of fat people' (quoted in Kirkland 2008: 180 n 20). Advancing a Fat Rights perspective, Anna Kirkland analyses diverse legal claims raised by fat people who were discriminated against at work and in public places.

She demands special spatial arrangements for them, and the transformation of purportedly neutral spaces, which humiliate and exclude fat people from the opportunities every citizen should enjoy. Fat people may diverge from the corporeal norm but, she argues, the norm is unjust and naturalises social and cultural privileges granted to thinner bodies (Kirkland 2008: 45).

Disciplinary bioascesis

Beyond claims about the obesity epidemic, the imperative of health and the continuous ascetic surveillance necessary to attain and maintain the highly demanding ideals of healthism oblige enormous discipline. The combination between the narcissism of a hedonist consumerist society and the commandments of bodily discipline generate ambivalence. Crawford's interviews of the early 1980s with Chicago-area middle class residents demonstrate the extent to which self-control and self-denial had by then become indispensable to health (Crawford 1984, 2006). However, as he later noted, discipline and self-control do not necessarily imply consumerism. From the 1990s onwards, at a time when the US became the country of hyper-consumption, middle class Americans experienced a conflict between discipline and pleasure. Health became the arena where the contradictions between control and release, denial and pleasure were to be tamed and resolved (Crawford 2006: 413). The expansion of knowledge about health hazards, media hype about potential health risks and a voracious consumption of medical information exacerbated the very insecurities they were supposed to mitigate, and generated a spiralling interaction between control and anxiety (Crawford 2004). In this context, 'working on the self by working on the body's health cannot deliver the symbolic assurance needed to offset either anxieties about the dangers of a toxic society or the deepening insecurities of contemporary American life' (Crawford 2006: 416; see also Courtine 1993).

Neoliberal subjects are called to take part in society simultaneously as self-controlled individuals and as hyper-consumers – even if, to achieve the difficult balance between eating and thinness one has to, like the bulimic, throw up what one eats. The perfect citizen is capable of spending more while wanting less, and the thinness cult seems to be the perfect place to achieve such a magical gesture; as Julie Guthman notes, 'spending money on becoming thin is the perfect solution for both neoliberal subjectivity and neoliberal capitalism more broadly' (Guthman 2009: 193). The eating-less-while-consuming-more dictum steered consumerist consumption towards the commodification of health, fitness and diet products; ambivalence in that context takes the form of repressing desires that might hinder the quest for health and corporeal perfection (Crawford 1984). Yet individuals who impose on themselves a rigorous discipline during the week often let go at the weekend, and commit forbidden excesses. Bioascesis and overkill are the two sides of the same coin. Similarly, the proliferation of risky habits and forms of living must be understood in the framework of a culture dominated by the analysis and avoidance of risk (Le Breton 2000, 2002b; Race 2009a). The healthiest obsession comes hand in hand with extreme cases of sedentariness, fast-food consumption and synthetic drugs. Musculated bodies coexist with chain smokers and self-satisfied fat people. Psychiatrists, psychoanalysts and psychotherapists attest to the existence, in numerous individuals, of a stressful and anxiogenic tension between bioascetic and permissive habits and ideals (Costa 2004; Orbach 2009).

Bioasectic discipline is offered as the demanding path to become 'fit' and live according to prevalent ideals of bodily perfection. Fitness is marketed as the panacea that will liberate us from medicine, a protection against the ills of modern society, the recipe for happiness and even fidelity ('couples who exercise together stay together', writes Chris Pepper Shipman in *I'll Meet You at the Finish!*), and the possibility of developing a coherent life in times of moral disorder and social disintegration (Glassner 1989: 183).

As cultural historian Hillel Schwartz writes, 'no diet comes without a larger social agenda' (quoted in Gusfield 1992: 75). In the 19th century, phrenology was credited with the power of contributing to general good health, and an avalanche of phreno-physiological literature sustained the belief in such a power (van Wyhe 2004). This literature had explicit moral aims. For example, by showing the pernicious effects of alcohol and sexual depravity upon the brain, it encouraged temperance and sexual moderation as rational prescriptions for a healthy life. Natural laws of health converged entirely with social norms, and the achievement of good health depended on following the organic laws that governed both body and brain (Cooter 1984; van Wyhe 2004). One author who made a major contribution to the popularisation of the phreno-physiological literature was Sylvester Graham (1794–1851), one of the founders of the natural food movement in the US, who opposed city life and industrialisation while underlining the virtues of agricultural, traditional life (Nissenbaum 1980; Sokolow 1983). In Graham's view, improving individual health enhanced moral capacities and vice versa. The coincidence of the moral with the medical spheres was a 19th century commonplace, according to which the self-discipline and self-control required to follow a healthy diet and hygienic principles were acts of moral excellence. One of Graham's main followers was the physician and Seventh Day Adventist John Harvey Kellogg (1853–1943), a prolific writer who continued the Grahamites' crusade for natural food and sexual purity, and in that context invented corn flakes (Carson 1957).

Joseph Gusfield (1992) has analysed the relationship between food and morality in the natural food movements from Graham and the Grahamists in the 1830s to the natural food movements of the 1950s–1990s. In the 19th century, self-discipline and self-control regarding a healthy diet were presented as acts of moral excellence and as a way to preserve social and moral orders that were perceived as threatened. Both the Grahamites of the

1830s and their descendants of the 1980s and 1990s are characterised by extreme individualism and the idealisation of an original natural state. Health and personal safety – understood as independence from the established order of social institutions – depend on a capacity for control which in turn requires a code of discipline as severe and demanding as the hedonist ideals it opposes. In the 1830s the crusade for self-control was supposed to counteract the weakening of family, communal and religious bonds, and counterbalance the rise of an economy that furthered the liberation of appetites and desires.

Graham lamented that the market replaced the home, and the responsibility for edibles production was passed on to food technology and commercial agriculture. He firmly believed that those economic developments engendered an individual lacking self-control. Hence, as Gusfield (1992: 89) writes: 'In a changed world of weakened tradition and institutional forces, it is to the self that one must look for sources of social and personal control'. Willpower, as embodied in ascetic sexual and alimentary practices, was supposed to allow the individual to regain control over the physical, social and political bodies.

From the point of view of the individual advice they deliver, health food movements in the US in the second half of the 20th century share with the Grahamists a stress on individual self-control, self-help and self-improvement, as well as scepticism towards social institutions. The 1990s movements, however, seem apolitical in comparison with their 1960s counterparts, which placed dietary advice in a more general ecological perspective and defined it in opposition to mainstream American culture and politics (ibid: 95). The prescriptions of the 1990s natural food movements were aimed at the independent consumer in a free market, and therefore fully embodied the ideology of bioascesis. Instead of a health and food counterculture, they envision

> . . . a world of individuals who carve out for themselves areas of detachment from the strong and overpowering institutions that surround them. Only in themselves and in their control over their lives and their intake can they find the health and the political morality with which to prevent disease and illness.
>
> (Gunsfield 1992: 99)

The bioascetic obsession with the body, its shape and performance has replaced the Grahamists' aspiration to reinstate a divinely ordained moral order. In the first decade of the 21st century, the relation between health and the commodified body remains as strong as it was in the 1980s and 1990s (Lupton 2003). It is within the somatic culture of healthism and bioascesis that the body becomes the matrix of personal identity, and selfhood is redescribed in corporeal terms.

The somaticisation of subjectivity

Nikolas Rose has convincingly argued that, in liberal societies of the Western world, at least during the six or seven first decades of the 20th century individuals talked, thought and acted upon themselves and others as if they were inhabited by a 'deep interior psychological space'. People related to themselves by means of a *psy* discourse and a vocabulary of unconscious desires, trauma, anxieties, neurosis and sexuality (Rose 2007: 188, 222). In such a cultural landscape, selves

> . . . were characterized by a profound inwardness: conduct, belief, value, and speech were to be interrogated and rendered explicable in terms of an understanding of an inner space that gave them form, within which they were, literally, embodied, within us as corporeal beings.
>
> (Rose 1998: 4)

Rose traces the ways in which this 'deep' psychological space began to flatten out in the last decades of the 20th century. The 'psy complex' has not definitely been replaced; rather, it coexists more or less harmoniously with new ways of acting, thinking and speaking about ourselves as corporeal, somatic selves.

As a consequence, psychological notions of personhood have been partly displaced by somatic/cerebral individualities. Cerebral identities are privileged forms of somatic selfhood, and these important mutations in subjectivity must be understood within the context of the spread of neuroscientific claims beyond the laboratory. Neuroscientific theories, practices, technologies and therapies influence the ways we think about ourselves and relate to others, favouring forms of neurological or cerebral subjectification. Thus, the idea of a 'neurobiologization of selfhood', expressed in the notion of 'neurochemical self' (Rose 2007) or the 'cerebralization' of the self, largely represented by the notion of the 'cerebral subject' (Ehrenberg 2004; Ortega and Vidal 2011; Vidal 2009), may constitute an underlying presupposition of the 'new brain sciences' (Abi-Rached 2008; Abi-Rached and Rose 2010; Rose and Abi-Rached 2013). Beliefs, desires, behaviour and emotions are addressed in cerebral and neurochemical terms, and their social and cultural instantiations and effects are also attributed to the brain (Rees and Rose 2004). The term 'cerebral subject' is a useful way to describe brain-centred approaches to the human person and to understand the use of brain-related terminology, and is in no way meant to reify the notion.

The cerebral subject is an 'anthropological figure' (Vidal 2009) that has no reality prior to its performative embodiments. It therefore points to processes of individuation and subjectification in specific socio-historical contexts. It coexists with ways of conceiving oneself and prolongs in radical form the earlier Western manners of constituting naturalised and singularised individuals (Dumont 1986; Elias 1978, 1982; Foucault 1978, 1986, 1990; McFarlane 1992; Taylor 1989).

Psychological, cerebral and neurochemical selves emerge in particular socio-cultural contexts as part of a broader 'genealogy of subjectification' in Western culture (Rose 1988: 23). Hence, the process of subjectification has ontological pre-eminence, and that is why, to analyse the cerebral subject, one should focus on its formation and the practices of self-constitution through which individuals fashion themselves in cerebral terms (Vidal and Ortega 2011). This process can be understood as involving 'technologies of the self' (Foucault 1988) and the penetration of expert knowledge in popular culture. 'Making up people' (Hacking 2002: 111) entails the creation of descriptive or diagnostic categories through expert knowledge; individuals assimilate these categories into their descriptions and practices of the self, and thereby transform them and bring about realities that experts must in turn confront (Hacking 1995).

Alcoholism, and more specifically what sociologist Scott Vrecko (2006: 301) defines as an emerging 'folk neurology of alcoholism', constitutes an interesting illustration of the tendency to map personhood and illnesses upon the body or brain. According to Alcoholics Anonymous's 12 step programme, to attain permanent abstinence from alcohol involves a far-reaching inner transformation. Alcoholism is regarded as deriving from a flaw of the self, which the individual must recognise and work upon through introspection and discipline (Valverde 1998). By contrast, neuroscientific models of alcoholism focus on the brain's neurochemical balance. Drinking activates the brain's pleasure circuits, but in some cases, the brain rewards drinking too strongly, generating intense cravings. The treatment then targets an over-stimulated endorphin system. Alcoholic Anonymous's complex and ritualised confessional practices are replaced by simple decisions about taking or refusing medication. In one addiction-recovery website analysed by Vrecko, a woman named Sophia described herself as loving 'the phrase "endorphin challenged"' because, she said, 'that is exactly what I am' (2006: 302).

Thus people take up scientific knowledge and descriptions and incorporate them in their self-understanding as members of a neurobiological human kind. This does not mean that neuroscientific explanations of addiction have entirely replaced older notions. In her qualitative interviews with individuals being treated with the psychopharmaceutical buprenorphine for opioid dependence, Julie Netherland observes that patients frequently incorporate a neurochemical understanding of addiction into self-accounts which give a central role to discipline, responsibility and autonomy, and emphasise the role of willpower in attaining lifelong abstinence. Addiction understood as a brain flaw increases, rather than diminishes individual responsibility. Keeping a normal and healthy brain implies a 'way of life characterized by autonomous, responsible citizenship', whose attainment requires not only medication, but also willpower (Netherland 2011: 172). Neuroscience thus emerges as one among several frameworks individuals have at their disposal to understand themselves and their relations to others (Netherland 2011;

O'Connor and Joffe 2013; Ortega and Choudhury 2011; Pickersgill, Cunningham-Burley and Martin 2011; Singh 2012b).

Autism provides another interesting instance of self-understanding in biological/cerebral terms. It constitutes a 'problematic' psychiatric category (Rosenberg 2006), whose ontological status is in dispute; it is the object of public negotiation by medical and non-medical actors. For some people, autism is a disorder; for others, it is an example of the brain's natural diversity. Research about a very vocal group of autistic self-advocates who call themselves 'neurodivergents' illustrates this heterogeneity (Ortega 2009a; Ortega and Choudhury 2011). As highlighted by the very notion of 'neurodiversity', autistics' identitarian assertion is often linked to the cerebralisation of their condition. People diagnosed with autism fashion themselves positively as 'autistics' and not as 'persons with autism' because their brains are 'wired' in an atypical way. Their identitarian affirmation is frequently associated with a rejection of psychological explanations and psychotherapies, which they oppose to their 'neurological self-awareness' (Singer 1999). While such self-awareness is largely constructed and explored online, thus allowing autistic groups to bypass the medical establishment, the preference for cerebral explanations depends in part on the availability of neuroscientific knowledge-claims beyond the laboratory.

An often-imprecise neuroscientistic vocabulary serves to construe autism as a positive attribute, and to stress the natural difference relative to non-autistic, 'neurotypical' experience and identity. At the same time, the neurodiversity movement exhibits a cohabitation of everyday ontologies. Individuals shift registers, such that 'my brain' may in their language serve to designate 'my mind' or, perhaps more precisely, just 'I' or 'me'. Cerebral metaphors and metonymies contribute to give the brain the kind of psychological depth usually, or formerly, attributed to the mind. The interpenetrations of mind and brain at work in the neurodiversity context reflect the persistence of important continuities behind updated vocabularies, and this is typical of the entire neurocultural world (Ortega and Vidal 2011).

The constitution of neurological identities is emblematic of the somaticisation of subjectivity linked to the spread of neuroscientific discourses. Recalling the distinction mentioned above, we may say that the processes just sketched in relation to alcoholism and autism take place at the molecular rather than at the molar level (Rose 2007). Yet the molecular level often mingles with the molar one of dieting, fitness and healthism. An interesting example of such blend (as well as of the cohabitation of ontologies) is the burgeoning industry of neurobics, or what I have called 'neuroascesis', that is, discourses and practices aimed at acting upon the brain so as to maximise its performance.[11] Hence, neuroascesis constitutes a specific form of bioascesis.

In the neurobics industry, the ideal of 'brain jogging' simply translates into training the mind; the mind is supposed to be what the brain does, but it is through the mind that neurobics seeks to improve the brain. Behind its neuroscientific varnish, the neurobics market that has been growing since the 1990s does little more than carry on, in updated form, self-help regimens that sometimes date back to John Harvey Kellogg's late 19th century hygiene and diet prescriptions. Moreover, neuroascesis discourses and practices draw heavily on metaphors and analogies from body fitness. Even the titles of cerebral self-help manuals often refer to 'neurobics' or 'brain gymnastics'.[12]

In several books and programmes the brain is depicted as a sort of muscle. 'Just as weight lifting repetitions in the gym or jogging strengthen certain muscle groups, mental exercises appear to strengthen and enhance cognitive functions over time' (Tannen [undated]). The manuals speak of a 'brain gym' and 'mental weight lifting' (CBS 2006). Train your 'cerebral muscles' (Goldberg 2001: 255), but in such a way as to avoid 'brain cramps' (Chafetz 1992: 72). Practise regularly the 'brain stretches' that will help you to 'burn some synaptic calories' and prevent you from becoming a 'mental couch potato' (Parlette 1997: 16); face the problematic fact that mental muscles can enjoy television, that true 'bubble-gum for the brain' (ibid: 152–53). Neuroascetical authors draw from sports in order to distinguish among levels of brain accomplishment or mental prowess: 'you do not have to attain the brain equivalents of Steffi Graf's or Michael Jordan's level of physical fitness to be quicker in conversation, better at solving problems, have richer memories, and livelier associations' (Chafetz 1992: 23). For 'those of you who wish to exercise your brain systematically as an athlete would exercise various muscle groups', manuals provide a well ordered cerebral training programme, and recommend hiring a cerebral 'marathon trainer' and keeping 'brain workout diaries' (ibid: 213–14). The vocabulary of bodily fitness is thus extrapolated to the brain itself. Bodily and cerebral fitness go hand in hand. Discourses of the cerebral subject thus transpose the vocabulary of fitness from the body to the brain.

In cerebral self-help, psychological and neurological ontologies of the self exist together, and the main topics of traditional self-help combine neuroscientistic jargon with the slang of body gyms. Neuroscientific advances, even those connected to cerebral plasticity, which have come to play a central role in contemporary neuroascetical discourse (Malabou 2008; Pitts-Taylor 2010; Rees 2010), obviously do not suffice to account for such phenomenon. Rather, the syncretism characteristic of neuroascesis is best seen in an ontological perspective, as having to do with the development of views about the human and forms of sociality and subjectification that involve notions and practices of the self and its relationships with one's own body and other people. In short, neuroascetical practices are tools whereby persons constitute themselves as somatic individuals, and more specifically as cerebral subjects.

What is it like to be a somatic individual?

But, one may ask, what is it like to be a somatic individual? What are the subjective experiences of persons who describe themselves in somatic/cerebral terms? Nikolas Rose and other scholars focusing on governmentality have distanced themselves from sociologists such as Anthony Giddens and Ulrich Beck (Beck 1992; Beck, Giddens and Lash 1994; Giddens 1991), who depict modernity as a process of increasing reflexivity and individuation (Rose 2007: 25).[13] For these sociologists, modern individualism involves an existential anxiety and a feeling of insecurity which are exacerbated by the discourses of risk promoted by the mass media and health promotion strategists. In contrast, governmentality scholars argue that self-anxiety emerges in particular socio-historical contexts. They reject sociological notions such as Giddens's 'ontological security' or 'practical consciousness' (Giddens 1991), which seem to them to constitute technologies of government, historical formations to be examined (Rose 1999; Vazquez Garcia 2005). For them, there is no general form of the subject, no single process of subjectification characteristic of any given historical period, age, culture or even individual. As Dean (1996: 218) put it:

> . . . the problem with contemporary social accounts is that they are pitched at too general level and propose a mysterious, even occult, relation between general social processes and events (e.g. globalization, postcolonialism, the emergence of postmodernity) and features of identity (fragmentation, fluidity, multiplicity, the desire for self-actualization, etc).

Scholars working in the governmentality framework notice that there is a multiplicity of presents, as well as a multiplicity of selves and ways of experiencing those presents. There is no unique way of being a somatic individual; rather, persons constitute themselves as 'somatic' while they pursue different goals, and experience somaticisation in diverse ways according to a variety of factors, ranging from individual physical and psychological make-up, race, gender, class and economic differences to the particular exposure to the cultural and economic demands of dieting, fitness and medicine consumption. And yet, when we use terms like somatic individuality, biosociality or cerebral subjects, we assume that, in any given socio-cultural context, there are socio-cultural forces or trends that shape individuals' beliefs and manners of being. If this were not the case, we could hardly speak of shared beliefs about normality, body shape or fitness. Pursuing a genealogy of subjectification involves identifying the forms of individuality that prevail in a given socio-cultural context.

In advanced liberal societies, somatic forms of personhood are favoured and coexist with other forms of self-understanding. Frequently using qualitative research, scholars from medical sociology, anthropology, philosophy and public health have examined common elements of the subjective dimension of somaticisation of selfhood.[14] To begin with, Anthony Giddens's observation in *The Transformation of Intimacy* (1992) that the body has turned into a 'visible carrier of self-identity and is increasingly integrated into life-style decisions, which an individual makes' has indeed become a truism (Giddens 1992: 31; see also Bunton 1997; Nettleton 1997). For many people, the body represents personal freedom, and the best of themselves. As a woman interviewed by the Brazilian historian Denise Sant'Anna (2001: 69) put it:

> My body corresponds to what I like, to what I am, independently of my genetic inheritance, my cultural roots, my social class, my civil status, and how I make my living. My home has my face, my bathtub and my clothes express what I am.

This statement, illustrative of middle und upper class Brazilian women, but not different from what many women of the same class in any developed country would say, identifies body and self.[15] Beyond cosmetics and fashion, bioascetic practices blend body and mind, such that personal identity is 'worn' on the body (Lupton 1995). For some scholars, the ascetic work on the body displays the inner truth of the self (Heyes 2007: 22). Bioidentities, however, are not physical manifestations of such inner truth, but rather, expressions of an identification of self and body. Somatic culture is a culture of total visibility where self and appearance coincide. This has important consequences for subjective experiences.

There is a major risk in turning the body into an *alter ego*. In contemporary somatic cultures, appearance becomes essence; therefore, what persons are is constantly under the scrutiny of others. Individuals are to a certain extent condemned to appearance, and deprived of the capacity to simulate, to hide feelings and intentions, to have secrets – a capacity that was present in the culture of intimacy, and which somatic cultures render superfluous. If we are what we seem to be, then only the gaze of others makes us real. Superficiality, suspicion, affectation and insecurity are features of the somatic individual, in whom control and regulation substitute reciprocity and openness to the other:

> The constricted and fragile self, through a psychocultural obsession with finding and maintaining identity, mounts a defense, one for which the borders are insecure. Thus, if in my insecurity, you try to speak your different truth to me, I will not hear you. What I will hear instead is the threat of incursion. How I perceive your personhood will be the distorted mirror of my own repressed self or a difference I am unable to integrate into my understanding. In such ways the other is irrevocably bound to the self; and since not admitted to a place of connection – through

> commonality *and* difference – the other will inevitably be relegated to a place of subordination, containment and control.
>
> (Crawford 1994: 1364)

The only way of escaping the tyranny of appearance is by conforming to the norm, by becoming so similar to others that one disappears from their view. Susie Orbach points to such a situation, but draws consequences opposite to mine. She argues that, although the call for bodily beauty and the goal of bodily perfectibility have been democratised, the democratic ideal has not generated aesthetic variation, but actually reduced it (Orbach 2009: 3). The existence of a small number of body types does away with 'body variety' (ibid: 14). The result is a 'homogenous visual culture' that provokes constant anxiety and vigilance: the 'norm has become to worry' (ibid: 4). While such homogenisation indeed generates anxiety and shame, individuals may seek it precisely because identification with the norm protects them from constant scrutiny, despite the fact that this protection is in itself selfish and contributes to the further stigmatisation of those who do not fit within the norm.

The (anti-)politics of bioascesis

Beyond the subjective impacts of somatic culture, in which paradoxically normalisation protects individuals from the constant exposure to the gaze of the other, the identification of identity and appearance tends to foster depoliticisation and weakens ethical standards. Llewellyn Negrin condemns the investments in the body as the only 'authentic' locus of self-identity: permanently to try to fasten one's identity to the body through practices such as body modification involves neglecting the changing constitution of the self (Negrin 2008: 30). Against postmodern assumptions that identity is reduced to the external bodily appearance, identity and appearance, although interrelated, are not synonymous and there is ethical value in insisting on the irreducibility of the former to the latter. That is why she argues that the aesthetic cult of the body and health 'results in a diminished conception of identity centered on the physical transformation of the self rather than on other forms of self-realization that could contribute to the betterment of society as well as an enrichment of the self' (ibid: 32). Healthism too promotes depoliticisation, exclusion and social conformity (Crawford 1980; Heyes 2007). Yet at the same time, as already mentioned, the conflation of identity and normalised appearance may protect from constant exposure to the gaze of others.

It would be nonetheless exaggerated to claim that somatic culture has led to the total privatisation and depoliticisation of the body. As we will see in the next section, practices of body modification may resist bodily normalisation of healthism and body consumerism. Additionally, the following chapter will examine the socio-political dimension of somatic culture, as well as the politicisation of the body in biological citizenship and different forms of health and disability activism. Those phenomena differ in goals and ideology from the culture of healthism analysed in this chapter, in that they are largely motivated by political and ethical goals, and offer innovative ways of thinking politics and political action (Epstein 2008; Hess et al 2008; Orsini 2008, 2009; Orsini and Smith 2010). In contrasting the ethico-political dimension of biological citizenship and disability activism with the apolitical character of healthism and body consumerism I do not wish to establish a harsh dichotomy. The mutations in the notions of citizenship (Ong 1999, 2006a, 2006b), from the political to the biopolitical, already presuppose the blurring of the public-private divide. Health movements and diverse forms of biological and disabled activism analysed in the next chapter collapsed that divide by turning corporal features into the basis for rights, inclusion, entitlements and protection.

In this context, it is useful to recall the notion of 'intimate public sphere', advanced in the late 1990s by Lauren Berlant to describe the political achievement of the conservative Reagan revolution since the 1970s. As she argues in her book *The Queen of America Goes to Washington City*, Reaganism succeeded in redirecting political energies 'into the sentimental spaces of an amorphous opinion culture, characterized by strong patriotic identification mixed with feelings of practical political powerlessness' (Berlant 1997: 3). The turn towards the family runs counter to discourses and practices of non-familial sexuality, as well as other forms of racial and economic difference. The American child and the American foetus become ideal citizens of the intimate public sphere. It is no surprise that in 2007 President Bush recognised Julie Aigner-Clark, the stay-at-home mother who created the Baby Einstein franchise of baby products, as the new face of American citizenship.

Against such a socio-political background it may be argued that the private bodies of healthism and body commodification have become public. Naturally, healthism and the pursuit of fitness and health involve forms of sociality and community. Friends and couples often train together, and gather for sport and health activities. Social relations are often integral to the fitness environment, and turn out to be significant determinants of the participant's subjective wellbeing (McAuley et al 2000). Furthermore, public health discourses' promotion of individual responsibility vis-à-vis one's own health are linked to the claim that, by taking care of themselves, individuals indirectly contribute to society at large: they save health care resources, which can then be spent elsewhere. Hence, the transformation of the public sphere into an intimate public sphere gives visibility to private and apolitical bodies – private and apolitical insofar as individuals are more focused on attaining perfect health and bodies than on the amelioration of society. However, the result is an 'anti-political politics' – in Berlant's words, 'a politics that abjures politics, made on behalf of a private life protected from the harsh realities of power' (Berlant 1997: 11). The same conservative ideology that turned public bodies into public emblems of American infantilised citizenship removed sex from the public sphere and directed it back to heterosexual bedrooms, enriched the foetus with citizenship rights, banished to the margins of citizenship immigrants, youth, black people, women and homosexuals and, finally, reproduced and strengthened gender, racial and economic violence.

Despite some resemblances, the intimate public sphere of contemporary culture of healthism and bioascesis departs radically from the ideals of classical ascesis. Greco-Roman ascesis had, as we saw, a fundamental socio-political dimension since its ultimate objects were the city and other individuals. As Peter Brown and others have insisted (Brown 1978, 1982; Clark 1999; Rapp 1999; Rousseau 1999), the ascetics of late antiquity embodied both the solidarity of the community and its values, needs, hopes and fears. Ascesis was an effective form of cultural and political resistance, an expression of alterity and of the will to constitute alternative forms of subjectivity and sociability. In the modern practices of the self or bioascesis explored here in connection with healthism and body consumerism, ascesis frequently becomes discipline and serves the anti-political politics of a public intimate sphere where diversity is transformed into a conglomerate of apolitical bioidentities.

Towards an embodied phenomenology of body modification

I would like to finish this chapter with an illustration of contemporary ascetic practices which, at the phenomenological level, may constitute a bulwark against the drive towards body normalisation operated in healthism and body consumerism. I refer to practices of body modification understood as forms of 'somaesthetics'. Philosopher Richard Shusterman advanced the notion of 'somaesthetics' to address the 'critical, ameliorative study of one's body as a locus of sensory-aesthetic appreciation (aesthesis) and creative self-fashioning' (Shusterman 2000: 138). He highlights the significance of somaesthetics for political philosophy. Cressida Heyes (2007) believes in its value as a tool of political resistance to normalised gendered bodies. Body modification practices may be taken as an example of somaesthetics in the sense of Shusterman and Heyes, that is, as a way of working on the embodied self without capitulating to normalisation. Obviously, tattoos and piercing, among other body modification techniques, are widespread, and have to a certain extent been commodified and turned into body consumerism. It would be naïve to say that dieting and cosmetics are per se repressive and normalising while body modification is per se transgressive. However, body modification practices frequently resist the ideology of identity and beauty and enable transformative experiences of embodied subjectivity (Heyes 2007: 127). In my view, they may represent an important instance of reappraising the body as a 'living dimension of individual experience and action', against its reduction, within the cultures of healthism and body consumerism, to a 'mechanical instrument of atomized parts and measurable surfaces' (Shusterman 2000: 161).

Body modification practices constitute special and *sui generis* attempts at personalising the body and thereby developing bioidentities. They may seem to follow an identitarian and apolitical pattern akin to that of healthism. However, from the standpoint of individual experience, they may also be seen as a way of escaping the culture of appearance and of recovering the connection between self and the lived body. I shall here explore this potential of body modification to oppose bodily normalisation, with an emphasis on how it brings about the phenomenological presence of embodied and material subjectivity.

Body modification encompasses a very broad range of practices. They include tattooing, piercing, branding, cutting and subcutaneous implants, but extend well beyond the skin to include body-building, fitness and wellness training, anorexia and fasting, as well as all sorts of prostheses, and nanotechnological and other internal devices aimed at enhancing or substituting bodily organs (Featherstone 2000). Here, however, I will focus on the first group (tattooing etc), as well as on some prosthetic and technological body modification systems. While visible tattoos and piercings have become commonplace, genital piercing in men and women, branding, burning, cutting and subcutaneous implants remain non-mainstream. There is a gradation in these practices, whereby the bodily mark becomes deeper and less epidermic. While some practices have been incorporated into consumer society and now serve cosmetic or ornamental purposes, others evolved against commodification and normalisation. The commercialisation of body modification has always been an important issue for body-modifiers, since it constantly obliges them to redefine their practices. In her ethnographies of body modification subcultures, Victoria Pitts observes that body-modifiers frequently defend the political and personal significance of their practices in 'producing new modes of embodiment that push the limits of normative aesthetics and often link pain and pleasure' (Pitts 2003: 12).

Bodily modifications have been approached in two main ways, both of which leave out significant aspects of the phenomenon. The first considers them as a constitutive element of consumer society, spectacle and fashion. The second identifies them with self-mutilation and regards them as mental health problems. Against these interpretations, I will advance an embodied phenomenological reading of body modification, as a form of somaesthetics, that is, a 'reflective, corporeal practice aimed at somatic self-improvement' (Shusterman 2003: 114), and a resistance to corporal normalisation. Before turning to the phenomenological reading, let us examine the two main approaches to body modification.

According to one approach, tattooing and piercing are no longer associated with subcultures or marginal groups, but reflect superficial trends, the process whereby exotic elements become part of the eclecticism and self-referentiality of contemporary fashion. Contrasting bodily marks in traditional societies with their incorporation in the so-called neo-tribalism or modern primitivism movements, British sociologist Bryan Turner identifies the latter with growing individualism. The individual he characterises as the postemotional actor of contemporary body modification practices is a 'member of the airport departure lounge, in the sense that he is blasé, indifferent to traditional signs of commitment and remote from the conventional signs of care' (Turner 2000: 49). In such a framework, bodily marks are superficial, consumerist and narcissistic. In the 'supermarket of style' they represent, in Baudrillard's words, a 'carnival of signs' without related meaning, an 'eclectic mish-mash' of once powerful signifiers cut off from their original meaning (quoted in Sweetman 2000: 53; Turner 2000).

Although of course such a mish-mash exists, body modification practices are irreducible to it. The permanent nature of the marks, as well as the pain and planification they involve, are not consistent with the superficiality of merely trendy choices; rather, those marks may be defined as anti-fashion, insofar as their irreversibility is the antithesis of the way fashion works. The diverse forms of body modification escape the flow of the commodification of 'change' in advanced capitalism (Fisher 2002: 102). As American tattooist Don Ed Hardy put it, although the practices in question involve elements of fashion, the tattoo is 'on your body, it's permanent; you have to live with it; and it hurts' (quoted in Sweetman 2000: 72).

The second approach to body modification, which also seems to me inadequate, considers the phenomenon as a pathology, as a mental health problem patterned after self-mutilation, a behaviour about which there is ample medical literature and for which a range of therapies, from the pharmacological to the psychodynamic, has been proposed (Favazza 1996: 288–323). This literature considers self-mutilators as 'deeply disturbed individuals' motivated by 'various psychopathological reasons' (Pitts 2000: 299; see also Favazza 1996; Myers 1992; Pitts 2003; Sullivan 2002). Body-modification practices are identified with forms of self-harm and self-mutilation, not only because the body is represented as imperfect, dysfunctional or ill, but also because it is seen as the expression of a suffering self in need of treatment. For instance, the psychiatric label of 'delicate self-harm syndrome' has been applied to addictive and repetitive skin cutting. This condition is more prone to afflict women and is linked to sexual abuse, depression, borderline personality and even psychosis (Pitts 2003; Sullivan 2002). When applied to body modification, the medical label denies agency to the self-mutilators and discredits their motives: 'Even though body modification is undertaken to change appearance, and for self-definition, it turns out that self-definition is not a choice, but a desire, addiction, need or urge' (Pitts 2000: 296, 2003: 18).

The media reproduces the pathologising discourse. On 22 April 1998, for example, the *Washington Post* described body modification practices as 'methods of ruining one's life'. The article stated:

> There is alcohol, of course, but also marijuana and hashish and heroin and cocaine and LSD; amphetamines and methamphetamines, barbiturates and airplane glue, animal tranquilizers and Ecstasy. There are the aesthetic means of self-harm: tattooing, baby piercing, scarification, anorexia, bulimia. There is the outlaw life: gangs, guns, crimes, prison.
>
> (quoted in Pitts, 2003: 24)

Such narratives give rise to a mixture of fear, horror and disgust vis-à-vis body modification, which they integrate into a moralist discourse where it appears as a set of regressive or deviant behaviours. But moralist and normative condemnations forget that, as philosopher Bernard Andrieu (2003: 200) put it, 'subjective solutions to the question of personal identity find in the body an authentic materialization of the subject'. Cressida Heyes provides three reasons for rejecting the equation of body modification with mutilation. First, if self-mutilation involves physical harm, risk of complications or long-term limitations of physical capacity, there is no reason to believe that cosmetic surgery is less self-mutilating than body modification. Secondly, the label 'modification' or 'mutilation' presupposes mainstream corporeal norms, such that self-mutilation frequently is no more than a body-transforming behaviour that defies conventions of body shape and beauty. Finally, if the mutilated body were the subconscious manifestation of loathing for a traumatised body, then there is no reason to see yo-yo dieting or body-lifting as improvements of the subject's bodily image (Heyes 2002: 126–27; see also Pitts 2003).

In his ethnographic study of the Californian subculture of 'non-mainstream' body modification, the anthropologist James Myers (1992) concluded that the reasons for altering one's body are as diverse as the explanations given for it. And a more recent review of the relevant literature identified 10 broad motivations for getting tattooed and pierced, which were related to beauty; art and fashion; individuality; personal narrative; physical endurance; group affiliations and commitment; resistance; spirituality and cultural tradition; addiction; sex; and to nothing in particular (Wohlrab, Stahl and Kappeler 2007). While critics of the two approaches sketched above promote a political reading of body modification, I will here highlight its link to bioascesis and the formation of bioidentities in a more phenomenological perspective.

As already discussed, within somatic culture and healthism, responsible individuals define their lifestyles by the search for health, fitness and the avoidance of risk. Health status, bodily performance, and particular conditions serve as the basis of social and subjective life. For some, investing in the body is a response to the dissolution of social links, isolation and the loss of the collective values and meanings that structure an individual's symbolic world.

Traces in the body become substitutes for those lost realities. Tattoos, piercings and other marks constitute 'body projects' (Shilling 2012) aimed at constructing a coherent identity, at providing a permanent self in a world experienced as changing, contingent, uncertain. Speaking about his tattoos, an individual emphasised that 'you cannot runaway from them, you can't stop being a tattooed person' (quoted in Sweetman 2000: 71). Tattooing, branding or scarification provide a feeling of cultural and social stability, and are typically used 'in situations where people feel . . . [the] need to preserve their individual and social identities and to advertise . . . the would-be permanence of their allegiances, values and beliefs' (Polhemus and Proctor quoted in Sweetman 2000: 63). The bodily mark functions for them as a sort of 'self-signature' (Le Breton 2002a: 21) by means of which the subject asserts an identity different from the one society assigns him or her. To the extent that bodily appearance is essential for personal identity, body modifications represent a quest for authenticity, an attempt at giving one's essence a physical, stable location. Fastening identity to one's skin may constitute an attempt to neutralise the nihilism involved in the experimentation with different looks and the postmodern aestheticisation of the self. Yet the human self is by its nature changing; it requires both coherence and flexibility – neither the postmodern parade of empty marks, nor the definitive branding of self-identity in the flesh (Negrin 2002: 30–31).

Corporeal modifications, however, leave permanent or semi-permanent marks. Moreover, their production is an invasive and laborious process involving pain, blood, careful preparation and later care. After his first genital piercing, one body modifier explains, the wound

> . . . bled like mad the night I got it done. The anaesthetic wore off, and I woke up about three in the morning absolutely saturated in blood, all over the bed, I thought I was dying, you know, it was gushing out . . .
>
> (Quoted in Sweetman 2000: 61)

As such testimony suggests, body modifications, especially the more radical ones, can hardly be considered mere signs to be consumed in a 'supermarket of style'.

By means of bioascetic practices, individuals find their essence in corporeal appearance, a truth about themselves that society does not provide. One heavily tattooed individual relates for instance that through the practice of tattooing, he felt 'a better, more rounded, and fuller person . . . more in tune with the person [he] really thought [he] was' (quoted in Sweetman 2000: 68). Body marks localise in the body subjective criteria of self-assessment, and support a somatic self that gives rise to a sense of freedom, authenticity and autonomy. This freedom, however, entails the individual's extreme singularisation through a project that provides concrete form to the 'psychological reconquest of the body' (Andrieu 2002a: 391). Another heavily tattooed individual interviewed by Sweetman emphasises how body modification represents self-possession and self-creation when he states: 'it makes you feel individual . . . you know like, everyone's born with roughly the same bodies, but you've created yours in your own image [in line with] what your imagination wants your body to look like' (quoted in Sweetman 2000: 68).

Corporeal marks may therefore embody the illusion of rupture with the symbolic order, the dream of a totally naturalised essence. As the symbolic limits of the body are strained to the extreme of the body's real boundaries, a new relation with the real seems to emerge. Body modification provides a phenomenological awareness of the body, a material localisation of identity and meaning. What, however, might be its political relevance? To what extent do such bioascetic practices really escape the common language of consumption, flexibility and technological invention or intervention of consumer capitalism? How may they counter body consumerism and the loss of community bonds? Is there any real difference between the possessive individualistic character of body modification and the commodification of the body in advanced liberal societies?

These are not easy questions to answer. A good part of the difficulty resides in the fact that the phenomenological strength of body modifications coexists with their limited political impact. Body modifiers frequently celebrate disconnection, distinction and difference. This is largely the case of cyberpunk and modern primitivism, which embrace a 'body-in-isolation' through their emphasis on self-creation and self-fashioning (Pitts 2003: 195). In his ethnography of tattoo conventions, Paul Sweetman stresses body-modifiers' implicit rejection of group affiliations, and the personal nature of tattoos and piercings. Few interviewees linked their body modification practices to some form of membership in specific subcultures (Sweetman 2000: 66–67). The refusal to recognise social, political and economic ties is particularly strong in transhuman, posthuman and cyberpunk contexts animated by the vision of a future without ideologies or power issues. Performance artist Sterlac celebrates an individualistic rhetoric of body-customisation and self-evolution, as well as liberal notions of choice and individual freedom based, as we saw earlier, on the obsolescence of body materiality and the consequent disappearance of socio-political and economic constraints.

Modern primitivist practices have been criticised as inauthentic, an expression of middle class elitism and a way of disentangling tattooes from their working class associations. As Aidan Campbell puts it, modern primitivism represents an 'insincere way of making life in the social margins seem more attractive' (quoted in Pitts 2003: 128). However, as Pitts notes, for many body modifiers modern primitivism has a political meaning, expressing cultural dissent and challenging or decentring identities. Rather than mimicking traditional identities and life forms, it promotes practices of 'queering identity' (ibid: 132). Similarly, for cyberfeminist body-modifiers, technology may undermine the permanent and fixed character of identity. Cyberpunk body-modifiers and body artists highlight the political and subversive nature of their practices, through which they intend to resist the biocapitalistic colonisation of the high-tech body. By using medical technologies in unauthorised ways, they contest medical authority and monopoly over those technologies. Yet, body modification practices do not fully break with dominant ideologies and their body projects (Pitts 2003). The liberal sense of freedom of choice regarding identity haunts cyberpunk experimentation as much as mainstream cosmetic medicine. This largely limits the political efficacy of those practices, which is outweighed by their phenomenological meaning.

To grasp the phenomenological meaning of body modification we have to consider the broader socio-cultural and historical context of the medical visualisation of the body analysed in the previous chapter. As we saw, since the Renaissance there have been recurrent attempts at reducing the experience of the body (the body I *am*) to the objective, measurable and quantifiable body I *have*. In the vocabulary of possessive individualism, we are free as owners of a body (Duden 1997), and having a body is more important than being one. During the Vesalian revolution, anatomy patterned itself on a mechanical model of the living being, and hence anticipated the classical rationalist view of the body as a material machine. During the 18th century, that paradigm was extrapolated to other fields, such as physiognomy, which found in the corpse an adequate metaphor for describing the human body (Stafford 1991). Thus arose, in the 19th century, the naturalistic reality of the body as object of research and treatment. The body-object or physical body became the target of a biomedical model,[16] which, sustained by visualisation technologies, contributes to the disincarnation of subjectivity and to turning the body phenomenologically obsolete.

In such a historical framework, corporeal modifications may be understood as a reaction against the objectified body of contemporary technobiomedicine, and as a recovery of the body as subject of sensation, experience and locus of being-in-the-world. Body marks procure 'an extinct body perception' (Duden 1997: 2); they constitute a 'subjective investment of the self in its own materiality', such that they allow the subject to reincarnate his being (Andrieu 2002a: 42–43). In contrast with the cultures of fitness and healthism, where the great attention to the body coexists, as we saw, with great anxiety about it, body modification practices try to recapture the body's subjective and lived dimension. Voluntarily risky behaviours, from radical sports to unprotected sex practices such as 'bare-backing', with their accompanying experiences of speed, fear and pain, have become socially unacceptable (Andrieu 2002a; Le Breton 2000, 2002b; Race 2009a). But what they promise is precisely to save the body from the 'sensory monotony' of social norms, to intensify experience against the sensorial anesthesia of contemporary culture.

Pain is a crucial element of body modifications, a way of experiencing the body in a culture in which suffering and death are no longer recognised as parts of the human condition (Le Breton 1998, 2000). The negative perception of pain is reinforced by medicine, which does not approach it as an existential fact with cultural, historical and moral dimensions, but as a physiological or pathological phenomenon (Bending 2004; Bourke 2012; Morris 1993; Scarry 1987). In his 1975 *Medical Nemesis* Ivan Illich criticised the vision that came to dominate medicine since the early 19th century, according to which pain is a 'passive happening inflicted on helpless victims', to be avoided 'even at the cost of losing independence' (Illich 1975: 151–52). The authenticity of pain as a subjective investment in corporeal materiality involved in body modification is a response to such vision. Body marks, in short, aim to be ways of speaking 'in the name of pain', of admitting 'people in pain into the company of human beings with the assurance that human life will not suffer but prosper as a result' (Siebers 2010: 191).

The culture of sensory anesthesia favours the mechanistic model of the human body: it suffices to lubricate the machine to eliminate suffering and pain. Such biomechanical understanding is sustained by the visual paradigm in which the image gains the status of experience and is endowed with materiality. As we saw in the previous chapter, owing to the 19th century invention of new medical visualisation techniques, touch lost its epistemological, clinical and diagnostic significance, and was replaced by vision, which ranked as more objective, immediate and accurate (Reiser 1990). There is a direct relationship between the suspicion vis-à-vis touch, technobiomedicine and the depreciation of bodily experience. Vision is the least corporeal of the senses. The dominant visual paradigm and contemporary visualisation technologies offer images as if they were the represented things themselves; they thus give representations a materiality they actually do not possess. From a phenomenological perspective, body modification practices represent a rejection of the dominance of vision, as well as an assertion of the desire not to be treated as an image, but to recover sensory experience as a way to access the lived body and counter bodily normalisation.

****

This chapter compares and contrasts ancient and neoliberal ascetic discourses and practices. Its intention was not to establish a radical and artificial distinction between classical Greco-Roman and early Christian ascesis, on the one hand, and on the other, neoliberal bioascesis, with the related practices of healthism and body consumerism. I am aware of the simplifications involved in opposing these two forms of ascetical behaviour. They worked, however, as a means of calling attention to how the Foucauldian terminology of ascesis and technologies of the self is used to analyse neoliberal identity formations. The decontextualised usage of Foucault's vocabulary downplays the ethical and political meaning of ancient ascesis, which thereby loses its potential to inspire challenges to neoliberal selfhood. It comes as no surprise that such appropriations of Foucault often fail to criticise the growing interdependence of neoliberal somatic ethics and biocapitalism.

Ancient ascesis had an ethical and political dimension. One could argue that bioascesis practices also have an important ethical component – 'the ethical work that human beings do to themselves in the name of health, longevity, and their vital existence' (Rose 2007: 255). And yet, the ethics of ancient ascetical practices presupposed not only the self working on itself, but also concern for the *polis* and its improvement. To highlight that contrast, I have focused on healthism and body consumerism. They constitute privileged sites of deployment of bioascetic discourses and practices, and their apolitical and conformist character is manifest.

Nevertheless, I also wanted to look at more potentially political ascetical practices within neoliberal societies, and it is in that perspective that I examined the phenomena of body modification using a phenomenological framework. Body modification is an outstanding illustration of the ability of some contemporary ascetical practices to challenge corporeal normalisation. The next chapter will look more closely at diverse health movements and forms of biological and disabled activism which radically depart from the apolitical practices of neoliberal bioascesis that have been analysed in this chapter. In those contexts, important political and legal implications follow from making individuals' biological and corporeal status the ground for claiming rights, entitlements and different forms of protection and compensation.

Chapter 4

Bodies on trial

Law is uneasy with bodies that ooze or are leaky, especially those that are fat, distressed, sick, dying, addicted or appear impermanent.

(Fiona Campbell, *Contours of Ableism: The Production of Disability and Abledness*)

Bioidentities and biological citizenship

In the previous chapter I analysed the formation of bioidentities as the subjective pole of the 'somatic culture' or, more specifically, of biosociality. We saw that somatic individualities (Rose 2007) result from recent mutations in personhood associated with biomedical technologies and life sciences. Genetics, pharmacology, brain imaging and reproductive technologies are among the more important instances of this transformation. Genetics in particular has attained the attention of several scholars who described the multiple forms and contexts of a geneticisation of personhood (Rose 2007). Terms such as 'neurochemical self' (Rose 2007), 'cerebral subject' (Ortega and Vidal 2011; Vidal 2009) or 'pharmaceutical self' (Jenkins 2011) epitomise, as we saw, the description of personal identity in corporeal, bodily terms and relate to the diffusion of imaging technologies and a neuroscientific vocabulary among larger audiences.

The subjective dimension of bioidentity has been explored in the chapter on bioascesis (Chapter 3). However, the social dimension, taken up by several scholars, relates to issues of blood donation, blood transfusion, personal identity (Waldby et al 2004) and the formation of biological citizenship among people with Hepatitis C (Orsini 2008). Biosocial groupings (Rose and Novas 2005: 442) or collectives formed around a shared bioidentity (biological make-up, disease or impairment) are becoming the base for new forms of political and collective action with important legal implications. Paul Rabinow's original formulation of biosociality also addressed collective identity practices, especially that which linked to patient activism and identity formation (Rabinow 1996b: 102). Understood as the forging of collective identities around biomedical categories, biosociality has been widely used in the last decade (Rapp 1999; Whyte 2009).[1]

Important socio-political and legal implications arise from citizens framing their politics around their biological and corporeal status. Recent 'mutations in citizenship' (Ong 1999, 2006a, 2006b) have been identified as a consequence of the flows of markets, technologies and populations, thus challenging basic oppositions, such as that of a citizenship linked to a membership in a national state, countered with the condition of stateless, or a territorialised citizenship as opposed to deterritorialised human rights (Ong 2006a: 504). Rights, entitlements, state and territoriality, that which is traditionally related to citizenship, are being dissociated from each other and rearticulated along a universal criteria of human rights and neoliberal values of flexibility, mobility and entrepreneurialism. According to their national and cultural context, actors call upon a constellation of claims detached from territorialised notions of citizenship but based on biological make-up, patterns of mobility, individual market performance, use of technologies and cyberspace or even sheer survival in displaced camps. They can all become the basis for rights, entitlements and protection. Consequently, 'instead of all citizens enjoying a unified bundle of citizenship rights', in this new political landscape, 'heterogeneous populations claim diverse rights and benefits associated with citizenship, as well as a universalizing criteria of neoliberal norms or human rights' (Ong 2006a: 500). Global markets promote idealised values of flexibility, mobility and individual self-governing for citizens who are seen as economic agents in tension with more traditional and liberal ideals of citizenship. The deterritorialisation of citizenship and its merging with market-oriented norms and values has also created new arenas of political activism and citizenship formation, such as the streets as an arena for mobilisation of the poor in South Asia and Latin America; the internet, as a major vehicle for political mobilisation and activism in countries such as China or Iran (Ong 1999, 2005, 2006a, 2006b); and also for biopolitical activism and self-advocacy in advanced liberal societies (Novas 2006; Orsini 2008, 2009; Orsini and Smith 2010; Ortega 2009a).

Health-based claims are among the more important demands of citizenship in contemporary biomedicalised societies. In the last decade a plethora of terms such as 'medical citizenship' (Scheper-Hughes 2005: 145), 'sexual citizenship' (ibid: 145), 'biomedical citizenship' (Biehl 2004, see also Biehl 2006), 'therapeutic citizenship' (Nguyen 2005: 124–43), 'genetic citizenship' (Heath et al 2004; Rose 2007) or even 'neurochemical citizenship' (Rose 2007: 216–20) have emerged. They aim to apprehend the myriad of heterogeneous phenomena which includes organ trafficking in Brazil and India; anti-retroviral globalism; the global politics of HIV; AIDS activism in Brazil; the transformation of high levels of certain genetic illnesses in potential sources of biovalue; parent groups campaigning for the recognition of Attention Deficit Hyperactivity Disorder (ADHD) and its psychopharmacological treatments in the UK and the US. These examples give an overview of the plurality of issues and cultural contexts in which these forms of activism take place.

Among recent transformations in the notion of citizenship, the term 'biological citizenship' has gained special attention by scholars working in the field of social health movements and is generally used to describe new forms of political activism and citizenship organised around a certain disease or biological trait.[2] Although there are already legions of publications dealing with the subject, it is essential to understand the original context in which the notion of 'biological citizenship' was first introduced, since it has important consequences for evaluating forms of biocitizenship in advanced liberal societies. In her seminal book *Life Exposed: Biological Citizens after Chernobyl* from 2002, anthropologist Adriana Petryna coined the term 'biological citizenship'. What struck her is how in the aftermath of the disaster in Ukraine, where an emerging democracy was deeply connected to a rough market economy, the 'damaged biology' of an entire population constituted the ground for 'social membership and the basis for staking citizens' claims' (Petryna 2002: 5). Petryna defined biological citizenship as 'a massive demand for but selective access to a form of social welfare based on medical, scientific and legal criteria that both acknowledge biological injury and compensate for it' (ibid: 6). In the particular socio-political situation in Ukraine, characterised by unemployment, inflation, corruption and degradation of legal and political categories, the very notion of citizenship was entrusted with the 'superadded burden of survival' (Petryna 2002: 7).

'After the explosion of the reactor', Petryna writes, 'an important parcel of the impoverished Ukrainian population negotiated social and legal inclusion in the most rudimentary life-and-death terms' (ibid). This meant categorically that people were divided into the biosocial categories of 'disabled', 'sufferer' and 'non-sufferer', according to a descending hierarchy of entitlements. In this perverse 'political economy of Chernobyl-related illnesses', the sicker the individual the more social protection he or she is granted. As a distortion of the welfare state where sick and disabled people receive social compensation, health care and other benefits because they are not able to work, in Ukraine where there were nearly no jobs outside the contaminated area, practically only biological citizens were granted any right or financial compensation. In this political economy, life and death became a form of currency through which people negotiated some form of financial compensation and state protection and individuals paradoxically gained a sort of citizenship to the extent that they were reduced to a 'bare life' (Agamben 1998) or, in the words of one of Chernobyl's maintenance workers interviewed by Petryna, to a 'living death', or that of a 'walking corpse' (Petryna 2002: 3).

A radical reconfiguration of the concept of citizenship had taken place in Ukraine. Traditional citizenship, understood as a birthright linked to territoriality, was unveiled as a clear disadvantage given the health and environmental hazards. When citizens no longer have state protection and are no longer endowed with national and legal rights, they are turned into biological citizens granted with human rights. In post-socialist Ukraine, socio-economic exclusion and biological survival coexist uneasily with the human rights discourse. Human rights are the only rights granted to biological citizens and, as Hannah Arendt (1979: 299, 300) reminds us, 'the world found nothing sacred in the abstract nakedness of being human':

> If a human being loses his political status, he should, according to the implication of the inborn and inalienable rights of man, come under exactly the situation for which the declaration of such general rights provided. Actually the opposite was the case. It seems that a man who is nothing but a man has lost the very qualities which make it possible for other people to treat him as a fellow-man.

Petryna traces mutations in citizenship and the emergence of biological citizenship in a social landscape of generalised corruption and corrosion of political and legal categories. In these particular circumstances, individuals negotiate their damaged biology as a way of obtaining some sort of 'pseudo-citizenship' granted with human rights and market compensations. Biological citizens constitute in Ukraine the failure of classical citizenship. In the Arendtian framework, a biological citizen is stateless, a dis-citizen, a mere human being who 'lost his place in a community, his political status in the struggle of his time, and the legal personality which makes his actions and part of his destiny a consistent whole' (Arendt 1979: 301). Nevertheless, he was better off than just 'citizens' in this post-socialist republic in transition to market capitalism.

Biological citizenship in advanced liberal democracies

The concept of biological citizenship was taken up by diverse scholars after Petryna and adapted to the context of advanced liberal democracies. Nikolas Rose and Carlos Novas have proposed a general definition of the term that might account for the different case studies and particular cultural contexts and which includes all those 'citizenship projects that have linked their conceptions of citizens to beliefs about the biological existence of human beings, as individuals, as families and lineages, as communities, as population and races and as a species' (Rose and Novas 2005: 440; see also Rose 2007: 132). With its introduction describing diverse social health movements and biomedical activism in neoliberal democracies, biological citizenship attained positive appraisal and, in some cases, even a romantic allure that was absent in Petryna's account. It catalysed mutations that depart from a classical Marshallian formulation of citizenship as a progressive development of civic, political and social rights (Marshall 1950). Biological citizenship embodies new forms of collective action and democratic participation, shifting boundaries between state and society, private and public, and is generally endowed with an emancipatory potential to contest the biomedical authority and reframe the role of citizens in political processes. Further, it generates new objects of contestation, new forums for debate, new issues for democracy and new styles of political activism (Rose 2007: 136–37).

Biological citizenship has been used to address a plethora of health issues, social health movements and health activism in very different national contexts, ranging from autism activism in the US and Canada (Orsini 2009; Orsini and Smith 2010) to people living with Hepatitis C in Canada (Orsini 2008); from AIDS activism in Brazil (Biehl 2004) to *in vitro* fertilisation technologies in Ecuador (Roberts 2008); from embryonic stem cell insertion in India (Bharadwaj 2008) to parents of children with genetic disorders in New Zealand (Fitzgerald 2008); from Alzheimer life-writing (Burke, 2008) to the European Patients' Forum and the European Public Health Association (Flear 2008); from mothers of children with genetic disorders using the internet in the US (Schaffer, Kuczynski and Skinner 2008) to AIDS biopolitics in Uruguay (Reihling 2010); from Fat Rights (Schessler-Jandreau 2008) to the Deaf community (Friedner 2010) and genetic databases in the UK (Petersen 2005).

Although Rose and Novas explore the notion largely at a conceptual level, hence their discussion has been criticised as programmatic and decontextualised (Whyte 2009: 11),[3] they highlight different forms of biological citizenship according to national contexts, biopolitical histories, notions of citizenship, modes of government and activism and through articulating political, legal and ethical claims. They remind us how differently biological citizenship is mobilised in Africa or South Asia compared with London, Paris or New York (Rose 2007). This is not just a question of being attentive to empirical and contextual analysis but of keeping in mind how, in some socio-political situations, biological citizenship is linked to the failure of social and political citizenships and rights. This happens to be the case in post-socialist countries (such as the Ukraine) and other developing countries. Conversely, in neoliberal democracies, where the concept has been widely used, it is connected to economic visions of citizenship driven by market-based values and ideals. Notions such as 'entrepreneurial citizenship', or 'technopreneurial citizenship' (Ong 2005: 343–46) or the more general term 'consumer citizenship' (Clarke et al 2007) summarise this form of citizenship. They are based on neoliberal values of flexibility, mobility and entrepreneurialism, in which individuals are treated as economic agents who make choices based on market and consumer culture. For some people, such as the Consumer Citizenship Network, an EU-funded thematic network of researchers and civil-society organisations from 30 European countries, consumer citizenship optimistically involves an 'individual who makes choices based on ethical, social, economic and ecological considerations' and who contributes actively 'to the maintenance of just and sustainable development by caring and acting responsibly on family, national and global levels' (CCN 2005: 7; see also Schrader 2007). For more sceptical scholars consumerism replaced citizenship, a community of citizens turned into a marketplace of consumers, destabilising democratic achievements based on equal rights and social justice and dominated by the so-called 'cash nexus'; that is, the exchange of money for the desired goods and services (Brodie 2002; Clarke et al 2007; Ong 2005).

Consumer-citizens exercise their choice of public services as they do in the consumer society. Their equality is based on the capacity of all individuals being 'free to choose' in the context of neoliberal inducements for individuals, families and communities to take responsibility for their lives, health, wealth and security (Miller and Rose 2008). Consumer-citizens and biological citizens are similarly active and responsible citizens, counterpoised to the 'passive citizen' of the social state as a citizen of rights and duties (Rose 1999). Biological consumers have the obligation of conducting their lives responsibly in relation to others, to live their lives through acts of calculation and choice (Novas 2006; Rose 1999, 2007; Rose and Novas 2005). Several critics have denounced the notion of biological citizenship, as a particular form of consumer citizenship, for being a 'dangerous alternative' to traditional political and social citizenship, for 'celebrating the market and individual consumers' and endangering 'representative democracy and social welfare' (Cooter 2008: 1725, see also Cooter and Stein 2010).

From post-socialist Ukraine to advanced liberal democracies, biological citizenship underwent important mutations: from designating the failure of a system in granting state protection and financial compensation to their citizens it became identified with active and responsible consumers in a market society. Whereas in the former Ukraine biological citizenship seems to supplant traditional territorialised citizenship, in advanced liberal democracies it can challenge or intersect with social and political citizenship. This fact is important for the argument I want to develop in this chapter, hence I fully agree with Wehling (2011), who points to the different possible negotiations around the notion of biological citizenship. Whilst for previously marginalised individuals in advanced liberal societies biological citizenship may have positive outcomes, it has the potential also to further stigmatise and exclude other individuals and groups.

Disability and dis-citizenship

Several scholars draw on the notion of dis-citizenship to address the dilemma of unequal citizenship, lived by many individuals with disabilities and other minority groups in contemporary societies (Devlin and Pothier 2006). Despite liberal emphasis on equality, the American legal system is grounded on ascriptive traits such as race, ethnicity, gender and disability. They help to determine the frontiers of citizenship and rights bearing and to build and maintain a particular national identity (Carey 2009). Notions of dis-citizenship or disabling citizenship stress the fact that several groups are excluded from formal and/or substantive citizenship and can only aim at attaining 'citizenship by instalments' in a 'hit-and-miss-affair' (Prince 2008: 33). To the extent that disabling citizenship constitutes a form of biological citizenship, if genuine citizenship includes a capacity for productivity, disabled individuals are not able to match the standards of production and are therefore denied full citizenship. The ideals of the active and consumer citizen – an entrepreneur of him or herself, who conducts his or her life 'as a kind of enterprise, seeking to enhance and capitalize on existence itself through calculated acts and investments' (Rose 1999: 164) – cannot be attained by people with disabilities (Goggin and Newell 2005).

The ideology of efficiency and productivity implied in active and consumer citizenship has been criticised as a product of an 'ableist' discourse[4] that assigns people with disabilities an inferior socio-economic and political status (Devlin and Pothier 2006). In liberal societies promotion of justice and equality implies 'the application of the same principles of evaluation and distribution to all persons regardless of their particular social position or background', thereby ignoring gender, racial and sexual differences and issues of disability (Young 2007: 79). In fact, the 'original position' of Rawls's theory of justice presupposes individuals entering the social contract as 'free, equal and independent', in Locke's formulation, possessing approximately the same amount of physical and mental capacity (Rawls 1971). They are all aware that their 'various native endowments such as strength and intelligence' all reside 'within the normal range' (Rawls 1993: 25; see also Nussbaum 2006: 108). Disabilities are put aside, for they may hinder people from 'being cooperating members of society in the usual sense' (Rawls 1993: 20). Only 'citizens regarded as free and equal and as normal and fully cooperating members of society over a complete life' (ibid) constitute the contracting parties. They alone are endowed with full citizenship and governed by the basic political principles exposed in the theory of justice. Obviously this idealised situation is limited to the very few and its range of acceptance does not accommodate people with cognitive and physical disabilities. The asymmetry of power between the parties hinders people with disabilities from being included in the social contract as fully equal partners (Nussbaum 2006, 2010).

Although acknowledged as a 'pressing practical question', Rawls postpones the interests of people with disabilities to a later legislative stage, since they are not 'fully cooperating' members of society. Disability will be handled as a matter of charity and not of basic justice (Nussbaum 2006: 110, 123). Individuals with disabilities ('persons distant from us whose fate arouses pity and anxiety', Rawls 1971: 84) are for him objects of compassion and may distract us from making moral judgments (Malhotra 2006). Moreover, the original position requires that the contracting parties, blindfolded by the so-called 'veil of ignorance', are deprived of any concrete information about their own and another's race, sex, class, status, ability, fortune or intelligence, whilst at the same time presupposing that these natural endowments reside within the range of normality. Liberal equality demands, by definition, that subjects are featureless, divorced from all external contexts so that they are defined as morally indistinguishable and formally equal. Liberal legalism neglects both people's socio-cultural context and their embodied experiences. Rawls's persons are pure wills without bodies or context (Halewood 1996: 1346–47). The subject of liberal philosophy of justice is not a real life person, a 'concrete other', but a 'generalised other' member of abstract humanity. The 'temporary and heuristic destruction of identity' in the original position turns 'concrete people into generalized legal subjects' (Douzinas and Gearey 2005: 127). The legal person is 'disembodied, genderless, a strangely mutilated person' (ibid), an empty vessel or cipher that hinders the recognition of his or her concrete and unique characteristics (Douzinas 2008: 41).

Such a difference-blind paradigm has been severely criticised by social movements and scholars committed to a politics of difference. Iris Marion Young (2007: 80) has rightly argued that 'identifying equality with equal treatment ignores deep material differences in social position, division of labour, socialized capacities, normalized standards and ways of living that continue to disadvantage members of historically excluded groups'. Besides, despite the emphasis on justice, within the parameters of current social distribution where people are treated as appropriators and consumers, all aspects of the social bond are regarded as commodities to be owned and exchanged (Douzinas and Gearey 2005). Owing to this double moral standard of the theory of justice, that, on the one hand, precludes any serious acknowledgement of difference and the situation of the 'concrete other', whilst, on the other hand, validates the current economic and social status quo – any explanation of substantial equality must take into account the material situation of the contracting parties and the moral standpoint of the 'concrete other' (Benhabib 1992). The 'veil of ignorance ought to be pierced' to allow the knowledge of the contracting parties' identities, including the existence of disabilities. Therefore, one person's disability status cannot be factored out and has to be incorporated when developing a theory of justice. This shift would grant disabled people full citizen status.

Liberal theories of justice and disability studies

These questions have been keenly discussed within the literature of disability studies and theories of justice. Disability scholars aim to incorporate the issue of disability in the original position of a theory of justice; giving people with cognitive and physical disabilities political and civil rights on a basis of real equality.[5]

Liberal ideology and liberal theories of justice are principally at odds with disability rights. Liberal models of personhood – understood as independent, rational and self-sufficient – and of society as an association of independent and equal individuals, cannot be matched by individuals with disabilities. Patterns of efficiency and productivity demanded by free and responsible consumer-citizens, models of reciprocal cooperation and equal sharing of benefits and burdens, are far beyond the reach of disabled persons. As disability studies scholar Michel Bérubé (2010: 100) rightly asserts, any performance criterion whatsoever, be it independency, rationality or productivity, 'will leave some mother's child behind. It will create a residuum of the abject, a fraction of the human family that is to be left out of the accounting'.

Furthermore, individualistic versions of personhood and liberal theories of justice often ignore two important issues. The first is that a large proportion of the population may encounter impaired physical and/or cognitive capacities at some point in their lives; a factor exacerbated by the likelihood of an increased lifespan, a circumstance found especially within our biomedicalised societies. *Pace* Rawls states that disabled persons are 'persons distant from us', whose destiny arouses 'pity and anxiety'; they can be any of us at a certain moment of our lives. Secondly, ideals of autonomy and independence do not only fail to apprehend and appreciate women's and disabled persons' experiences, but are also a fabrication.[6] In our highly complex societies everybody is dependent in one way or another. No society would work without an extreme level of dependence reaching all aspects of human living from mobility and communication to production of food. Ideals of independence are philosophically arbitrary, conceived relative to any given society's expectations about what people usually do for themselves and how they do it (Wendell 1996). Independence, according to Eva Kittay (2001: 570), is a fiction, whose pernicious effects 'are encouraged when we hide the ways in which our needs are met in relations of dependences', and that turn those who cannot mask their dependence into 'pariahs', or objects of 'disdain or pity', as in Rawls's theory of justice.

Besides, personal autonomy is not the opposite of expert governability, but one of its main tools. In advanced liberal societies, autonomous individuals are governed through their freedom. 'The autonomy of the subjects and targets of governments is not a threat' observes Nikolas Rose, who concludes that: 'autonomy can be allied with such objectives as economic success, national population policy, conceptions of the desirability of education and training and the like' (Rose 1999: 50). Liberal principles of liberty, autonomy and choice do not describe the reality of the majority of persons living with disabilities. Moreover, who wants a society whose ethical ideals are principally unattainable by a large percentage of its population? I can only agree with Susan Wendell (1996: 149) when she asks: 'should a society have ethical ideals that are universally applied but which some people are precluded from attaining because they have certain kinds of bodies?'

Hence, and following Young's suggestion, if disability is not an uncommon condition unrelated to our ideals of justice and personhood and, if independency is a social projection of an ableist society, we can learn about issues of social justice and, more specifically, about 'structural inequality, normalization and stigmatization', if we regard disability as '*paradigmatic* of structural injustice, instead of considering it exceptional' (Young 2007: 85). Following these insights a number of scholars, gathered in recent years around the emerging 'critical disability theory' (Pothier and Devlin 2006)[7] or 'disability legal studies' (Mor 2006), have tried to advance an agenda that places disability at the core of a theory of justice. Inspired by critical legal studies (CLS) and disability studies these new fields problematise the legal construction of disability and the power dynamics involved in it. Their agenda involves an anti-necessitarian understanding of disability that goes beyond abstract rights and focuses on the genuine inclusion of people with disabilities (Devlin and Pothier 2006). They also pursue a comprehension of citizenship and justice for disabled persons, in tension with consumer citizenship and liberal theories of justice, defined by fairness of treatment rather than equality of outcome (Turner 2006). Critical disability studies acknowledge the capacity of people with disabilities to 'fully participate in the original position as equal contracting parties' that can only be possible if 'there is an explicit commitment to eradicating both structural barriers that handicap people with disabilities and ableist attitudes that contribute to disability oppression' (Malhotra 2006: 79). The inclusion of disability at the core of the original position involves bringing to the centre of a political philosophy of justice what was traditionally relegated to the margins and abjected. It is a vehicle by which to provide justice to the singular and the universal, to address the fundamental paradox of justice; an impossible task, as Derrida (1992: 17) formulates: 'To address oneself to the other in the language of the other is, it seems, the condition of all possible justice'. However, it is 'not only impossible', but even 'excluded of justice as law'. Disability legal studies (DLS) and critical disability theory regard disability as a category that is socially constructed through power relations and apparatuses. Disability issues do not only refer to impairments, functional limitations or personal tragedy; they also concern social values, institutional and legal priorities and, above all, power. Any attempt to challenge them necessarily involves the shifting of liberal discourses of liberty, autonomy and choice, opening up a space for more contextual, power-related discussions. Law is the place where the 'meaning of disability is constantly formed and transformed, contested, negotiated, defied, interrogated, constrained and liberated' (Mor 2006: 78). DLS' critique involves going beyond contemporary legal scholarship on disability, which usually applies in instrumental form the research done in disability studies, to analyse court decisions or legal policies dealing with disability law. Current legal scholarship treats law, separated from society, as an instrument for social change, which ultimately consists of modifications of statutes, doctrines or policy directions within the legal realm. DLS aims to undermine the legal construction of disability itself, the role of law in producing and maintaining disablement and the cultural assumptions and images of disabilities informing its narratives and those informed by them (Campbell, F. 2009; Mor 2006).

Critical disability theory, in turn, aligns with critical race theorists and radical feminists in challenging the assumption that difference can be downplayed, as suggested by liberal theories of justice. Just as the Constitution cannot be colour or gender-blind, since overlooking race and gender issues perpetuates racism and patriarchy, to ignore the difference introduced by disability prolongs stigma and exclusion. Critical disability requires that differences be confronted, not by-passed or downplayed. Let us take the example of Canada's Supreme Court decision in *Granovski v Canada (Minister of Employment and Immigration)* [2000] that governments should not 'fail to recognize the added burdens which persons with disabilities may encounter in achieving self-fulfilment' in a society and a legal system 'oriented to the able-bodied' (*Granovski v Canada* 2000: 33).[8] In this legal context formal equality is unsatisfactory for disabled people since 'ableist norms that ignore difference' can make 'participation simply impossible' (Devlin and Pothier 2006: 12). The persistent presence of ableist institutions' beliefs and norms places persons with disabilities in an inferior position. As Justice Gérald la Forest in *Eldridge v British Columbia (Attorney General)* [1997] has stressed: 'disabled persons have not generally been afforded the "equal concern, respect and consideration"' demanded by section 15(1), the equality rights provision, of the Canadian Charter of Rights and Freedoms. Instead they suffer under 'paternalistic attitudes of pity and charity' and their participation in the social mainstream 'has been conditional upon their emulation of able-bodied norms' (*Eldridge v British Columbia* 1997: 56).[9]

Devlin and Pothier (2006: 18) summarise the goals of a critical disability theory as follows: first, to bring to the fore the unequal situation of people with disabilities; secondly, to undermine 'necessitarian assumptions' that reinforce stigma and exclusion of disabled persons and, thirdly, to help promote individual and collective agency of people with disabilities in their struggles for 'recognition and redistribution'.[10]

Disability rights activism and biological citizenship

I have written earlier in this chapter that biological citizenship introduces important mutations in traditional notions of citizenship. It has also been argued that in advanced liberal societies biological citizens share many features of consumer citizens, as active and responsible entrepreneurs of themselves, whose rights resemble the rights of consumers at the marketplace

(Rose 1999). Consumer citizenship exceeds and enables biological citizenship within neoliberal governmentality. While not everybody is a biological citizen, most of us have undergone this significant mutation in citizenship and have been compelled to activate the powers and rights of citizenship as consumers in a market society. Can we consider 'disability activism'[11] a form of biological citizenship? If, as we have seen, dis-citizenship or disabling citizenship highlights the exclusion of several groups from formal citizenship, then it can be understood as a form of biological citizenship. In this sense Eileen McDonagh (2009: 96), for instance, argues that biological citizenship encompasses disability as a way to integrate a bioidentity (defined by impairment) into the legal and political discourse. However, the incorporation of disabled identities into citizenship faces two important challenges which deserve to be closely analysed: (i) the limitations of the social model of disability and (ii) the Janus-faced nature of biological citizenship.

Limitations of the social model of disability

Disability rights movements and disability studies are committed to the social model of disability and the rejection of the medical model (for a canonical account see Oliver 1990; see also Wendell 1996; Shakespeare 2006). The social model establishes the distinction between impairment and disability. Whilst the first is a biological and individual fact, the latter is a social construction. People are disabled by society and its disabling barriers, not by their impairments; therefore, the primary political goal is the removal of barriers. If disability is a social construction resulting from the power structures of an ableist society that discriminates against disabled people, a better approach seems to be anti-discrimination campaigns and struggles for equal citizenship and civil rights, modelled on the American with Disabilities Act (ADA), approved by the US Government in 1990 (Fleischer and Zames 2001; O'Brien 2004), or British equal opportunities and race relations law, specifically the UK Disability Discrimination Act 1995 (Shakespeare 2006). In this context, social transformation and rights rather than cure or pity are the appropriate strategies. The claims of disabled people are often articulated within a vociferous anti-cure rhetoric, as illustrated in the case of autism activism (Chamak 2008; Orsini 2009; Orsini and Smith 2010; Ortega 2009a; Silverman 2008).

The social model has been criticised for downplaying the importance of impairment and for consequently equalising disabled and non-disabled people. Underlying the idea of a barrier-free world is the assumption that the problem is not the embodied experience of the individuals' impairments but the failure of society to accommodate them (see among others Shakespeare 2006; Wendell 1996). Yet, what is more important for the discussion in this chapter is that the social model is principally at odds with biological citizenship as it is usually conceptualised. If impairments are largely downplayed by disability rights movements, they (including also genetic status, biomedical diagnosis or nosological categories) are generally emphasised in issues surrounding identity formation and health-based claims within discussions around biological citizenship (Hughes 2009). They are different styles of political activism with distinctive and sometimes contradictory agendas. It is worthwhile to illustrate this discrepancy with two different and extreme styles of activism, as in the case of contested illnesses and autism activism.

Chronic fatigue syndrome (Bülow 2008) and excessive daytime sleepiness (Kroll-Smith 2003) are surrounded by uncertainty concerning their aetiology, treatment, prognosis and diagnostics. Their reality as uncontested nosological categories may be disputed, constituting what Joseph Dumit has called 'illnesses you have to fight to get' (Dumit 2006; see also Moss and Teghtsoonian 2008). They have names but no codes, and therefore are not included in the International Classification of Diseases (ICD). This is particularly relevant in the context of the US political economy of mental health care, where access to health care is increasingly restricted through bureaucratic procedures, and doctors' views about the status of particular illnesses are enmeshed with the decisions of employers, insurers and health maintenance organisations (Scott et al 2000). In this wider context, illnesses with no evidence of conclusive biomarkers or consensual objective criteria are 'contested', and individuals must fight to get them. We observe a form of biological citizenship articulated around the belief in the 'reality' of the impairment as a way to obtain full insurance coverage and other forms of compensation that depend on the acceptance by the specialist's community, the medical establishment and other stakeholders engaged in the process of determining whether the 'disorder' justifies the pleaded reparation. This recognition usually involves the search for a somatic or material cause of the disorder, for 'social legitimacy presupposes somatic identity', as historian Charles Rosenberg (2006: 414) has observed. In these cases, biological citizenship demands the recognition of the reality of the impairment, as the more important political strategy.

There is also an opposite trend to this form of biological citizenship, organised by the movement of patients and their families who struggle for recognition of the nosological status of certain diagnoses. It involves a different form of negotiation, as in the case of anorexia. On several 'pro-ana' websites anorexia is presented as a lifestyle choice and the medical view of the disorder is rejected. Pro-ana sites are more numerous and better organised than the anti-anorexic websites and have a higher number of visits (34,988 as compared with 27,878) (Clarke and Amerom 2007; see also Miah and Rich 2008). For some scholars, the pro-ana movement is closer to the disability rights movement in its belief in the social model and in its rejection of medical approaches to anorexia (Tierney 2001).

Autism activism also represents a very interesting case, since one finds both forms of negotiation within the biosocial field of autism research and advocacy. First, there is the biological citizenship model, personified by autism advocacy organisations which insist on the epidemic dimension of the disorder and challenge society and institutions to take autism research and treatments seriously. This model also includes the legal mobilisation of parents to secure funding for behavioural therapies for their children (Freeman 2003). Organisations such as the National Alliance for Autism Research (NAAR) and the Cure Autism Now Foundation (CAN)[12] bring together families, medical doctors and scientists devoted to accelerating the pace of biomedical research in autism and raising funds for research and education in the area. Therefore, their activities can be considered to invoke a form of biological citizenship as they gather together individuals diagnosed with autism, their families and caretakers and draw them into the activism surrounding the condition. They are also close to the biomedical model since people with autism are framed as passive recipients of social and medical services and not as active agents. As a result some scholars have explored to what extent it is possible to speak about citizenship when the individuals involved have their capacity for agency severely limited (Orsini 2009; Orsini and Smith 2010).

There is also another form of biological citizenship within autism activism which largely corresponds to the anti-vaccination movements. Those involved in these groups believe that there is an important link between autism, the MMR vaccine and other vaccines containing mercury (Blume 2006; Offit 2008). The issue of vaccine safety has been the subject of legal claims raised by parents who believe certain vaccines harmed their children (Orsini 2009). In this form of advocacy, biomedical expertise is contested rather than instrumentalised. Although lacking enough scientific credibility, anti-vaccination movements represent a form of biological citizenship which challenges research evidence with their own evidence about nature and the causes of autism (Offit 2008; Orsini and Smith 2010).

Finally, we find a form of activism which is aligned more closely with the social model of disability rights movements and their anti-cure rhetoric. This form, largely embraced by autistic self-advocates, defies biomedical definitions of the condition and considers autism a difference rather than a pathology. Autism is a constitutive part of their identity and therefore they strive to develop an autistic culture and society (Ortega 2009a; Silverman 2008). They criticise the first form of activism for demonising autistics, frightening their families, promoting narrow viewpoints on the disorder and never listening to the life experiences of autistic adults.[13] The growth of self-advocacy movements and their enhanced exposure in the media have intensified the political clash between autistic and pro-cure activists. Conversely, to those for whom autism is a disease (primarily caused by genetic and neurobiological factors), the anti-cure positions of several self-advocates not only contradict their claims, but also insult the suffering of patients and their families. However, for the most extreme self-advocates, the search for therapies illustrates denial and intolerance towards differences, as well as the promotion of eugenic and genocidal policies.[14] They resist the negative vision of autism represented in sites developed by the professionals and autistic children's families, whose 'obsession' with a cure is considered disrespectful of the autistic way of life. To the extent that autism is not considered an illness but a difference, the search for a cure becomes an attempt to erase this difference. This is why the anti-cure perspective has gathered strength within the autistic self-advocacy movement (Sinclair 2005).

Autistic self-advocates' adherence to the social model, which in some cases downplays the subjective or personal impact of the impairment in individuals with autism, is at odds with the usual conceptualisations of biological citizenship. These, as we have seen, often emphasise the role of the impairment in constructing and negotiating personal identity, health-based claims and compensations. Nevertheless, despite their differences in scope and goals, as well as in their commitments to nosology, treatment and policy strategies, scholars have discussed the three forms of autism activism within the framework of biological citizenship (Orsini 2009; Orsini and Smith 2010).

Let us now turn to the second challenge of incorporating disability activism within the framework of biological citizenship.

The Janus-faced nature of biological citizenship

If, as has been argued, consumer citizenship has the capacity to exceed and enable biological citizenship within advanced liberal governmentality, we should also consider the possibility that this development may, by extension, result in critical consequences for individuals with disabilities. For, although biological citizenship opens up a space for framing the disabled person's claims for equal rights and justice, it likewise has the potential to stigmatise and exclude them. Patterns of performance and efficiency are established that cannot be matched by disabled citizens. They can raise citizenship claims within the framework of consumer citizenships, but the same social context denies these rights and regards them as dis-citizens. Biological citizenship simultaneously enables and conceals citizenship endowments for disabled people. The same culture that establishes standards of productivity and efficiency in individuals with disabilities opens up the possibility for contesting those standards. Or, more precisely, because we live in the context of consumerism, where citizens' models of performance and efficiency cannot be attained by persons with disabilities, disabled people have to struggle for a recognition and justice that takes into account their differences. Is there a way out? Disabled people seem to face an impasse, as *qua* biological citizens they are generally granted only some form of dis-citizenship. In this sense, it is worth keeping in mind Michel Bérubé's (2010: 100) observation that the 'desire to include all humans in a theory of justice is hopelessly utopian. The United States . . . managed to pass a national disability rights law, but its courts rule against disabled plaintiffs more than 95 per cent of the time'.

I am not suggesting that persons with disabilities are unable to attain full citizenship because we live in advanced liberal societies where citizenship claims are increasingly framed in the context of market and consumer culture. The paradox is that it is precisely this socio-political context that enables their claims to full citizenship, even if they are most of the time only granted 'citizenship by instalments'. It is precisely within the somatic culture of consumer citizenship in which disabled people are more exposed and disqualified as dis-citizens that their claims to formal and/or substantive citizenship become more manifest.

Biological citizenship reproduces the paradox existing within somatic culture. In this sort of biosocial setting, as we have seen in the previous chapter, individuals seek to improve physical fitness, live longer and share their diagnosis with others. A medical-physicalist vocabulary based upon biological constants, cholesterol rates, muscle tone, physical performance and aerobic capacity gains, in that context, a quasi-moral connotation and also supplies criteria for individual assessment. An individual's moral aspirations are measured against his or her bodily performances. The same ableist culture that establishes ego ideals based on physical performance and obsession with bodily perfection paradoxically includes forms of corporeality and body-images that do not correspond (and/or are diametrically opposed) to these standards. The more we expand aesthetic demands and ideals on corporeality, the more we open up a space for persons who cannot match the same aesthetic and corporeal ideals. Within neoliberal and multicultural societies we find, on the one hand, the spectrum of those we are willing to accept has expanded to include individuals with 'minoritarian' identities; those perceived as unproductive, ugly, or disabled. Paradoxically, on the other hand, within the same socio-cultural context, ego ideals are found to be hugely demanding and individuals are exhorted to be beautiful, productive, competitive and healthy (Crawford 1980; Greenhalgh and Wessely 2004; Rose 1999; Skrabanek 1994). Yet, it is precisely within this somatic culture, with its tyranny of corporeal normativity and performance, that legislation such as the highly controversial 'Ugly laws' (Schweik 2009), which existed in early 20th century Chicago and which were used to prohibit disfigured people from entering public places, would seem horrendous and unacceptable to us today. Since obsession with normality and perfection is corrosive to our wellbeing – ultimately unattainable (Martz 2001) and also highly stigmatising to particular sections of society and other minoritarian identities – the requested critical strategy is twofold: first, to challenge those highly demanding ego ideals and, secondly, to accept and promote full inclusion of individuals with disabilities and other minoritarian identities.

Biological citizenship and human rights

Discussions around biological citizenship are frequently linked to debates on human rights. As we saw in the case of post-socialist Ukraine, human rights were the only rights settled for biological citizens in the absence of national and legal rights. Discourses on human rights are frequently called upon to restore citizenship to millions of people in larger parts of Africa, Latin America or Asia, as well as excluded populations of migrants in marginalised zones in the larger metropolis of the developing world. This biopolitical reframing of human rights discourse coexists alongside the dramatic growth of NGOs and the emergence of a 'humanitarian reason' (Fassin 2011), a related 'humanitarian industry' (Ong 2006a: 504) and a 'market for humanitarianism' (Rabinow 2005: 48).[15] From Arendt to Agamben and others, there has been a certain scepticism vis-à-vis the reappraisal of human 'bare life' as the basis of citizenship entailed in contemporary biopolitical discourses and practices of human rights. One can only agree with Ong (ibid), who observes: 'It is by no means clear that the right to survival will everywhere be translated into citizenship or merely legitimized on the grounds of common humanity'. It is not enough to invoke human rights, just like magic, to grant rights to those excluded or marginalised. Hannah Arendt was very aware of this when she wrote in the already quoted passage: 'The world found nothing sacred in the abstract nakedness of being human' (Arendt 1979: 299).

From a critical legal theory's standpoint Costas Douzinas has also strongly debunked human rights discourses and practices, in both their political and their psychoanalytical dimension. Politically, human rights have at the imperial level become the *lingua franca* of the neo-imperial state and the justification for a 'military humanism' (Douzinas 2008: 208). At a more domestic level human rights claims expose practices of exclusion, domination and marginalisation that pervade society, while at the same time hiding their origins by domesticating struggles and resistances in terms of 'legal and individual remedies', which may lead to 'small individual improvements' (ibid: 109). To invoke the language of rights does not abolish exclusion; it only modifies its appearance. Fundamentally, human rights are only an attestation of 'political impotence', for 'the more rights we have, the more insecure and unfree we feel' – the so-called 'Foucault's law' according to Douzinas (ibid: 50). However, human rights also attest to psychological impotence, which corresponds to the related psychoanalytical critique: since the expression 'I have a right to X' frequently corresponds to 'I desire or want X' or 'X should be given to me', this identification undermines human rights as 'central principles of political and legal organisation' and exposes the major role of the 'public recognition and satisfaction of individual desire' in Western capitalist societies (ibid: 12). Human rights are depoliticised through their identification with the pursuit of desire. The psychoanalytical dimension of human rights that acknowledges the constitutive role of desire undermines their political potential, despite the fact that, according to Douzinas, they are becoming the organising principle of Western democracies. In our 'rights culture' everything can be turned into a legal claim. As a consequence, 'endless desire' and 'escalating fears' can erode human relationships, fragment communities and are substituted by a 'collection of atomised beings defending themselves' (ibid: 50). Human rights are deeply paradoxical, for the more rights we have, the less protected we are, as exposed by the fact that their ideological consolidation coexists alongside their greatest violations.

Human rights and disability rights

Disability activists argue that social citizenship failed as an effective mechanism for the equal inclusion and participation of people with physical and intellectual disabilities[16] and therefore have frequently called upon human rights in their quest for action. As a consequence, various global discourses on disability rights have emerged, such as a number of UN declarations on the rights of individuals with disabilities.[17] Human rights would be based on notions of vulnerability and dignity shared by everybody, rather than duty or contributions restricted to able-bodied individuals (Turner 2006: 109). Owing to their potential for inclusion, human rights approaches are a powerful strategy for the claims of individuals with disabilities. Without rights, Carey (2009: 31) reminds us, disabled people 'have been subjected to euthanasia, institutionalization, and brutality against which they had little recourse'. Yet, human rights have limitations and have been therefore criticised in a similar vein by Douzinas and others. Critical disability theorists have emphasised the tension which exists between human rights and their promises of full inclusion and the discriminative practices that exclude people with disabilities. Hence their scepticism vis-à-vis abstract human rights discourses and their focus on authentic inclusiveness (Devlin and Pothier 2006).[18] More specifically, in the US, a traditional lack of respect for human rights hampers these liberties being established and enforced through legal and political institutions. In addition, human rights aim at achieving the minimal economic security and access to health care for everybody and, as such, would probably be considered insufficient for those disability rights activists who strive for greater levels of inclusion and participation. Lastly, the provision of such rights, as in education and health care, entails the development of institutions that examine the needs of disabled people and which provide access to resources when required. Whilst supporting citizens' rights, the creation of these institutions can also stigmatise and threaten larger inclusive objectives of human rights philosophy (Carey 2009).

Rights can be disempowering, masking oppression and causing stigma for individuals with disabilities by creating new technologies of power and control over them. Often inclusion goes hand in hand with conformity (Tremain 2005) and the inclusion rhetoric inherent in human rights can act as a more excluding mechanism than the exclusion itself. Rights can be part of a dispositive of normalisation by supporting disabled people who take part in public life in prescribed ways and, at the same time,

> . . . exerting control over most of this population by granting access only to supervised settings such as education, developing narrow legal applications, using ambiguous wording, instituting categorical requirements that divide the population and demanding compliance to prescribed norms prior to accessing rights.
>
> (ibid: 32)

For example, the Americans with Disabilities Act (ADA) of 1990 was presented as an equal opportunity law, which would grant rights to the disabled (Fleischer and Zames 2001). In this context, eligibility focused on whether the individual was disabled or not. To gain access to equal opportunity rights in employment, individuals with disabilities were forced to emphasise their impairment and difference from the able-bodied, thus reinforcing stereotypes that the ADA was supposed to confront (Carey 2009).

Furthermore, rights rhetoric is often not accompanied by an adequate enforcement of those rights, leading to an illusion of justice without a real commitment to its realisation. Government and policy-makers engage with different human rights documents on the rights of persons with disabilities, without defining concrete enforcement measures to implement them. Disability scholars consider that, without an explicit mechanism for their implementation, recognising human rights for individuals with disabilities does not make sense and can even be insulting. If we consider Bickenbach's typology of different basic expressions of human rights for disabled people, enforceable anti-discrimination legislation is more compelling than voluntary human rights manifestos. Yet, even anti-discrimination legislation, such as the ADA, has been criticised by disability scholars for reinforcing the medical model, for its ideological underpinnings that lead to define some disabled people as morally at fault and its frequently unfavourable court outcomes (Bickenbach 2001). Carey illustrates how, despite documents such as the 1975 Developmental Disabilities Assistance and Bill of Rights Act that granted the right to appropriate treatment in the least restrictive setting to all individuals with development disabilities, the US Supreme Court (*Pennhurst State School and Hospital v Halderman* 1981, III-B) determined that the Bill of Rights Act 'does no more than express a congressional preference for certain kinds of treatment'.[19] Analysing a number of recent decisions in Canadian courts about the education rights of individuals with disabilities, Malhotra and Hansen (2011) arrived at a similar disappointing conclusion: despite different national and international documents regarding the rights of persons with disabilities, the Canadian Supreme Court frequently does not consider international human rights law as related to disability.

In decisions specifically concerning education and health care access for children with disabilities, such as *Eaton v Brant County Board of Education*, *Auton v British Columbia* and *Wymber v Ontario*, the Supreme Court of Canada did not deem international human rights law to be relevant.

There is a huge gap between human rights legislation and manifestos and real commitment to their implementation. Inclusion seems to be reduced to mere rhetoric without being translated into concrete measures for individuals with disabilities. In the case of the US, the rights of persons with disabilities granted in the ADA of 1990 have frequently been subverted. Concerning employment issues, rigid standards of disability have been imposed that made many disabled people ineligible on the basis that their disability was not substantial enough or did not limit their major life functions (Carey 2009). We can understand the scepticism of some disability rights scholars vis-à-vis the ADA, when its interpretation by a 'business-friendly judiciary' allowed the employers to win 'over 95 per cent of the suits brought under the Act' (Bérubé 2009: 209). Cost-effectiveness turned out to be one of the main rationales for court decisions based on the ADA; hence, Michel Bérubé's (2009: 211) lucid observation:

> Of course, many of us would prefer to believe that our children have intrinsic human rights and human dignity no matter what; irrespective of any form of human social organization; regardless of whether they were born in twentieth-century Illinois or second-century Rome or seventh-century central Asia. But this is just a parent's – or a philosophical foundationalist's – wishful thinking. For what would it mean for Jamie to 'possess' rights that no one on earth recognized?

Destabilising legal personhood

Disability studies thus provide an important analytical tool to scrutinise human rights law and policies related to disability. But more generally, as I have already discussed in relation to the theories of justice, disability can undermine liberal legalism and liberal constructions of personhood. It can destabilise the political theoretical construction of the abstract disembodied subject of liberal political theory through a series of important shifts such as, first, challenging the ideology of efficiency, productivity, rational agency, autonomy and individualism and, secondly, bringing the living embodied person, his or her subjective experiences and socio-cultural context (the 'concrete other') to the core of philosophical legal theories and practices and finally debunking normalised ideals of corporeality and ableism. If disability undermines certain assumptions of legal personhood and liberal legalism, it can also be produced and perpetuated by legal discourses and practices. 'By legalizing ableism', writes Mor (2006: 78), 'the law becomes constitutive of disability in itself'. In the age of disability rights, the law has the power to produce disablement and to exclude, confine and determine rights, entitlements and duties. It can judge to whom it is providing or denying accessibility, what is or what is not reasonable accommodation and whether the claimant is 'disabled' or disabled enough. Hence, aligned with the agenda of critical legal studies and critical race theory (CRT), the role of critical disability theory and disability legal studies is to use analytical tools not only to evaluate court decisions or legal policies regarding the rights of individuals with disabilities but also to examine the legal construction of disability and ableism itself; in other words, to scrutinise the precise ways in which the meaning of disability is produced, perpetuated, ignored or contested in a legal setting.

In parallel with this line of argument, Fiona Kumari Campbell has argued convincingly that, whilst equalisation initiatives in legal contexts have granted benefits for many disabled individuals, the negative ontology of disability has so far not been challenged. Courts remain attached to the so-called 'personal tragedy model' of disability, according to which, disability is intrinsically intolerable and therefore '*inherently* negative' (Campbell 2005a: 109). The issue here is that the ontological assumption supports many juridically sanctioned discrimination claims by reason of disability, whilst conversely supporting compensatory initiatives and compulsion towards medical interventions. Disability studies contribute to the critiques of the normative citizen of neoliberal societies, those who are classed as white Anglo-Saxon Protestant, heterosexual and able-bodied, which incarnates legality. According to Campbell (2005a: 112), feminist and racial rebukes of alleged impartiality of the legal subject should also take into account the 'ableist underpinnings of that subject'. Legislation and case law define disability as a pre-given feature of human bodies and downplay the effect of disabling barriers upon the impairment. Recognition of disability within legal discourses involves subordination, normalisation and regulation. The experiences of disabled individuals are standardised within juridical discourse through the process of subjecting them to the normative or ableist narrative of disability; the 'personal tragedy' model. Law regarding disability produces a 'legal fiction' according to which an adequate handling – limitation, prevention, cure when possible – of disability requires the negative ontology of disability coupled with biomedical prescriptions. Campbell draws on different sources that describe disability as 'not a desirable state to be in', suggesting that action should be taken to 'prevent or limit' it (2005a: 118). Embedded within these assumptions are a number of significant issues that may have crucial socio-political and legal consequences and which, therefore, are worth addressing.

We already know that within advanced liberal societies, where ideals of efficiency, productivity and rational agency are extolled, individuals are called upon to exercise responsible citizenship by assuming control of their health and wellbeing. Yet ironically, within this very same context, where disability is deemed negative, the disabled person equally is called upon to take responsibility for his or her own health and wellbeing. Individuals with disabilities are expected to cooperate with treatments, surgery and other processes designed, in theory, to 'enhance' their bodies and bring them closer to the ideal model of the able-bodied. This process implies 'morphing ableism', that is, creating the illusion that the disabled body is transformed into a normal body (Campbell 2005a: 119).

Where we find disability rights activism and its related politics, the social model of disability frequently coexists alongside outspoken anti-cure rhetoric, as in the case of autistic self-advocacy, a matter already addressed in this chapter. The 'obsession' with a cure is perceived as an annihilation of the diversity of the human body and brain and suggests a lack of respect for the lived experience of the disabled person. In accordance with this agenda, if disability is culturally perpetuated then its cure and all the values related to it are also social constructions: 'If you do not believe there is a disability, if you do not believe there is anything that needs to be "cured" or genetically prevented – that disability is indeed little more than a social construction – then you will likewise be freed from the need for a cure' (Cheu 2004: 209).

Where there are issues surrounding the application of anti-discrimination legislation, some scholars have distinguished between 'immutable' disability and so-called 'elective' (voluntary) disability. The former applies to circumstances where disability cannot be eliminated – at least, not with the present degree of technological and biomedical development – whilst the latter alludes to conditions that were caused or even worsened owing to the voluntary conduct of the individual (Key 1996; Tucker 1998). Consider heavy smokers who develop lung cancer, Deaf people who reject surgery for a cochlear implantation, individuals with diabetes who refuse to take insulin or a person with manic depression who declines medication. A conscious decision to maintain the disability is an individual's prerogative; however, Key and Tucker argue that 'society should not have to pay for it'. The maxim 'reasonable accommodation' should not apply and protection by the ADA should not extend to those who are choosing elective disability. Tucker is more emphatic and, referring to Deaf people who refuse to have their deafness corrected in the name of a Deaf culture, argues that they lack the moral authority, in an age of budget crisis, to demand that other members of the community pay for expensive accommodations required for compensating their deafness (Campbell 2005a, 2005b; Tucker 1997, 1998).

According to Campbell arguments such as these 'are dangerous', for they provoke a rearrangement of disability that 'effectively casts bodies, which are corporeally anomalous, out into the wilderness'. In this legal model, where there is no place for 'subaltern or previously unarticulated experience of impairment/disability' and which departs from its biomedical definition, disabled people find themselves with a difficult dilemma: either to undergo the technological procedures and surgeries to cure their disability or remain without legal protection (Campbell 2005a: 121). Taking measures to mitigate disability can also become a trap, as in the frequently cited case of *Sutton v United Airlines Inc.* (1999), in which two twin sisters applied for positions as pilots with American Airlines. They were denied the positions, the argument being that they did not match the company's vision requirements of at least 20/100, even though, when they wore corrective lenses, their vision was similar to those employees without visual disability. The two sisters sued the company under the ADA, alleging disability discrimination. It is notable that the defenders argued that they were not disabled, since their vision was corrected using visual aids. The US Supreme Court settled that individuals who mitigate their impairments should have this factor considered when evaluating if they have right to coverage under the ADA.

I believe Campbell has every reason to debunk legal ableist narratives and their legal fictions about disability that couple together the negative ontology of disability with biomedical prescriptions. Aligned with critical disability theorists and disability legal studies' scholars she criticises the law's commitment to the social construction and cultural production of disability through the exclusion of anomalous corporealities not reducible to ableist normalisation and the hindrance of 'non-cooperative' individuals from accessing welfare and other compensation rights. She is also right in being suspicious of the unknown consequences of many technological procedures necessary to turn bodies 'nearly able'. In fact, as Tom Shakespeare observes, side effects and other costs of cure can be even worse than the impairment itself. This occurs in the case of people with mental illnesses who report frequent experiences of lethargy and other negative symptoms as a consequence of the long-term intake of pharmaceuticals. Other illustrations of this include the pain and infections resulting from limb-lengthening procedures, ear damage caused by cochlear implants and, in the case of limb transplants, the side effects caused by suppressive drugs (Campbell 2005a; Shakespeare 2006: 109).

Popular representations of cure are frequently accompanied by a rhetoric of hope and enthusiasm that does not correspond to the actual findings of research and is not followed by any real benefits for people suffering with these conditions, as gene therapy, xenotransplantation, stem-cell or neuroscientific research frequently shows (Leibing and Tournay 2010; Ortega and Vidal 2010; Shakespeare 2006). The publicity around this research frequently creates a dynamic of expectations that may have significant consequences for the patients and their families, by giving them false hopes and alienating them from society even further (Brown and Michael 2003; Smart 2003). Hence, many disability rights activists totally disagree with cultural representations of disability championed by celebrities such as Christopher Reeve, who almost exclusively campaigned for biomedical research with very little interest in disability rights. Reeve was criticised, as we saw in the previous chapter, for giving false hope, being unrealistic and acting in a way which did not correspond to the lives and experiences of many individuals with disabilities.[20]

Moreover, interventions designed to normalise or provide cosmetic improvements, such as plastic surgery for individuals with Down's syndrome, are considered denigrating by activists since they, by extension, promote individual correction and normalisation rather than attempting to remove social barriers. In addition, as the threshold of normality is continuously shifting and boundaries between enhancement and pathology are becoming fluid, higher and higher levels of cognition, attention and other intellectual performances are demanded from individuals in order to fulfil the demands of active, responsible citizenship. This shifting of boundaries creates new forms of disability and dis-citizens. The level of what is considered normal grows ever higher and therefore more difficult to attain for people with disabilities, so the able-bodied of today become the disabled-bodies of tomorrow. Let us take the example of height enhancement. As the population becomes taller shortness may be devalued, which could have important consequences for individuals who are considered short in height. Peter Conrad (2007) describes this situation as a form of 'heightism', where tall people are given more chances: they are more likely to be hired for a given job or be better paid. So, within this context, certain assumptions are made about competence based on height. These real or imagined scenarios create very real anxieties in parents about their children's height and, furthermore, influence their predisposition towards biomedical interventions, such as the supplying of human growth hormones. Countered against this position individuals with some form of restricted growth or dwarfism are more limited when it comes to biomedical intervention, as this would involve complex processes, which promise to augment their height, such as painful limb-lengthening procedures (Campbell, F. 2009; Shakespeare 2006).

I believe that Campbell and other scholars who take a similar stance are wrong for a number of reasons. First and more generally, there are huge varieties of impairment experiences among disabled individuals influencing different attitudes towards disability, cure and biomedical interventions. Positive self-identifications, identity claims regarding disability and rejection of cures are common among Deaf people, tetraplegics, individuals with restricted growth, the obese or individuals diagnosed with high-functioning forms of autism. On the other hand, the situation is very different for those suffering with spinal cord injury, low-functioning forms of autism, acute degenerative illnesses such as multiple sclerosis or disabilities resulting from the ageing process. There are therefore diverse attitudes vis-à-vis the impairment, which include negating it, developing a positive sense of self, considering it a pathology, seeking medical interventions or hoping for a cure. Hence, Shakespeare (2006) has every reason to state that people have different motivations for turning to medicine and different attitudes towards their impairment that go beyond simplistic oppositions.

Disabled individuals' accounts are produced from standardised social models of disability coupled with oppressive politics of identity which negate the multiplicities of their experiences. Radical interpretations of the social model downplay impairments and the related embodied experiences of disabled individuals. Such rigid ideology claims that disability has nothing to do with individual impairments and everything to do with social barriers. However, this stance has been criticised by disability scholars working within the framework of cognitive disabilities, such as Eva Kittay (2001), Michel Bérubé (2010) and others. They argue that it is easier to speak of disabling barriers when referring to wheelchair users than when referring to those suffering with major cognitive disabilities (Bérubé 2010). For radical advocates of the social model, impairment-specific responses or impairment-specific organisations and approaches are superfluous since disabled people share a *common* experience of oppression, regardless of the impairment. It is also superfluous to survey the impaired population, to discriminate against types of impairment and the different experiences of the people living with them. Disability is all about 'structural exclusion' and not individual differences (Shakespeare 2006: 31). Rachel Hurst goes so far as describing the focus on impairment as a form of 'disablism' (another word for 'ableism', and which is seen as on a par with terms such as racism, homophobia or sexism). According to her, 'to concentrate on the personal characteristics of the disabled individual and the functional limitations arising for impairment is itself, disablism' (quoted in ibid: 40).

Disability identity politics

There are also dangers involved in building an identity on victimhood and oppression resulting from disability identity politics. The victim position implied in both the medical and the social model of disability – people are either victims of their flawed bodies and brains or of an oppressive society – denies the agency of disabled people as well as a positive engagement with their impairments and society. Additionally, it leads to a litany of oppression and wretchedness, which precludes significant progress in disabled people's lives. The politics of identity fuelled by the ideology of victimhood results in 'exaggerating the differences and the polarity between the minority group and the mainstream', turning politics 'more extreme, separatist, vanguardist and aggressive' (Shakespeare 2006: 80).

'Passing' is the term used to describe those who hide their impairments or do not want to 'come out' as disabled. Identitarian ambivalences such as these and the different experiences of people living with some impairment are frequently ignored by radical activists within the disability movement and dismissed as internalised oppression or false consciousness (Shakespeare 2006). This perspective can be patronising and can discount the complexity involved in identity constitution, the management of disclosure and the complex process of coming out (Bagatell 2007; Davidson and Henderson 2010). As Simi Linton writes: 'both passing and overcoming take their toll' (2010: 230).

To focus narrowly on relations with people sharing the same condition excludes or limits other important domains of human society that constitute our daily lives. Human identity is constantly being created and recreated through interaction with other individuals. It is formed by an enormous variety of beliefs, traits and positions, the 'shifting articulation of all these disparate elements or "subject positions" which combine in various ways, occasionally and transiently under the direction of one particular dominant element, but other times without any particular hierarchy' (Douzinas and Gearey 2005: 194–95). It is important to listen to those who disagree with generalisations made by members of one particular category about the experiences of all members of that category (Wendell 1996). Thus, for some people, disability is an important part of their lives, but nonetheless just a part of their lives, which is simplified, essentialised and decontextualised by identity politics. As Susan White (2009: 119) has observed: 'it is, after all, a life and not an identity that people are usually seeking'.

Identity politics tend to reify identity, promoting identification with one condition or life aspect, 'a single, drastically simplified group-identity which denies the complexity of people's lives, the multiplicity of their identifications and the cross-pulls of their various affiliations'. It also hides intragroup struggles, reinforces intragroup domination and 'lends itself all too easily to repressive forms of communitarianism, promoting conformism, intolerance and patriarchalism' (Fraser 2000: 112).

Deafness mitigation

I believe that Campbell and other scholars also show a biased analysis of one of the most discussed issues within the debate around elective disability: the question of the correction of deafness. It is noteworthy that, for Deaf activists, rather than a disability, deafness is a cultural identity, seen even as constituting an ethnic minority, with a common language. They do not identify themselves with disabled individuals since they do not believe they have any impairment whatsoever. This debate on the mitigation of deafness basically revolves around issues of embryo selection and cochlear implant.[21]

The topic of deaf embryo selection was raised a decade ago when a deaf lesbian couple in the US deliberately tried to have a deaf child (McLellan 2002; Mundy 2002; Spriggs 2002). They sought a deaf sperm donor from a sperm bank in order to increase the chances of conceiving a deaf baby. Since congenital deafness is a condition that disqualifies potential sperm donors, they asked a deaf friend from a family of five generations of deafness and were successful. When the baby, Gauvin, was born in November 2001 his hearing loss was confirmed. The case caused controversy; however. the parents claimed not to be making 'a political statement about Deaf culture' (quoted in McLellan 2002: 1315). For the parents, trying to have a deaf baby was no different to attempting to select a baby girl or a baby with black skin, since all three, the deaf, females and blacks can be likewise discriminated against. Furthermore, according to Spriggs, although 'Gauvin's parents will not permit him to have [a hearing aid] at present. They say if he wants [one] later, they'll let him have one' (Spriggs 2002: 283).[22] Some Deaf associations have shown sympathy and strong support for the couple's decision. They criticise the medical model, according to which Deaf people have to be fixed and are subjected to the hearing culture through the implementation of hearing devices and embryo selection for hearing babies. They also blame the media's portrayals of Deaf people as 'abnormal, inferior, needing to be fixed in order to be successful in life' (quoted in McLellan 2002: 1315). The decision also provoked harsh disapproval. Nancy Rarus, from the National Association of the Deaf simply does not understand why 'anybody would want to bring a disabled child into the world', since they 'don't have as many choices'. Ken Connor, president of the Family Research Council, a pro-family organisation, went even further and considered that intentionally to produce a disability in a child – added to the 'disadvantages' resulting from 'being raised in a homosexual household' – is 'incredibly selfish' (quoted in Mundy 2002).

While one has to acknowledge the predominance of the medical model and its ableist underpinnings in public approaches as well as negative media portrayals around the topic, I believe the choice of deliberately selecting a deaf embryo is problematic, as I will show. Although the couple declared not to be making any political statement, their choice can only be understood against the background of a radical identity politics and militant 'Deaf Pride' (Padden and Humphries 1988, 2005; Sacks 1990). Against this backdrop of frequently mutual hostility between the deaf and the hearing world, dialogue is often difficult to establish and any statement or action coming from one of the cultures can often be seen with distrust and hostility from the other. To illustrate this point, only in such radicalised socio-cultural milieu can a variation of the traditional Cinderella story emerge, in which the idealised difference between those who are oral (Deaf people who were oralised) and those who are deaf, is stressed. According to the story, the Deaf Cinderella gets new gloves allowing her to sign and receive beautiful jewellery designed by Deaf artists. She goes to the party at the Deaf Club and falls in love with the Deaf prince. At the end of the story the Deaf girl becomes the Deaf princess and her gloves allow her to 'erase her many years as an oral person and to gain the difficult but admired skill of signing like a native' (Padden and Humphries 2010: 401).

The issue of embryo selection is in itself complicated. Disability rights movements have vehemently opposed embryo selection for impairments including deafness. They called this practice selective abortion or 'selective eugenics'. For them aborting a disabled embryo is immoral, a genocide of disabled people, and a clear message that it is better to die than to live with disabilities (Hubbard 2010; Saxton 2010). I fully agree with their arguments that no parent should be compelled to abort a foetus with any impairment against their will. However, if we accept the argument that it is immoral to make a selection to eliminate certain impairments, for the very same reason it must therefore also be immoral to make a selection that results in the presence of an impairment. One cannot argue that the former is 'selective eugenics' while the latter is just protecting an endangered culture. If it is allowed and sometimes encouraged to select deaf embryos, according to the assertion that Deaf people are a minority cultural group, it also has to be allowed that individuals should have the option to avoid deafness through embryo selection. Furthermore, since the latter diminishes the political power of Deaf culture, it may not be a good strategy for Deaf activists. When the same arguments can be used for and against selection of deafness, selection against deafness will probably outweigh selection in favour of it. In our ableist culture of consumer citizenship it is more likely that preferences for avoiding any impairment will outrank preferences for ensuring it. Also for technical reasons, while technologies for selecting any impairment will only be used in a small amount of cases, prenatal diagnosis as a means of avoiding deafness could be used in any pregnancy (Anstey 2002). Therefore, I argue that defending the selection of a Deaf embryo may result in worse political consequences for Deaf culture than not selecting.

Among the Deaf couple's reasons for choosing to have a deaf baby was that it is not wrong intentionally to try to have a baby who might experience harm when those harms are socially imposed. In addition, since blacks, women and the deaf are vulnerable to various forms of socially imposed harms there is basically no difference in deliberatively selecting a foetus that may fall into the category of one of these groups. On the one hand, it is understandable that parents want their children to share some fundamental component of their personal identity – even if it is an impairment, in this case deafness. However, it may be considered selfish of the parents deliberately to allow the child to suffer some form of harm, no matter if it is just or unjust, or whether it is socially or physically imposed. I certainly agree that children should have the 'right to an open future', that parents should contribute to 'reduce suffering and limitations of opportunity in the world' and, above all, to ensure avoidable harm. To say that the harm a deaf child experiences is socially imposed does not absolve the parents from the responsibility for those harms: whether they result from social discrimination or not – their children will probably suffer if they are deaf (Anstey 2002: 287, 288). I believe it is unjust not to protect one's children through ensuring avoidable harm, even if the rationale is that otherwise the children will not belong to one's culture (notwithstanding the fact that, as we will see below, hearing children can fully participate in that culture and also in the hearing world).

Therefore, to some extent I agree with Savulescu (2002) but would take issue with some of his points; for example, his argument that a couple has the right to procreate whomsoever they want but with the caveat that there are surely some exceptions to this right. He also observes that some Deaf people desire to have a Deaf child and take steps to enable it. However, I do not agree with his argument that there is no ethical issue in this. In my opinion, there is indeed an ethical issue since the couple knows, first, that there is a high probability that the child will experience harm because of the disability (whether socially produced or not) and, secondly, because in approving this practice they cannot consider it immoral for individuals to take measures to select non-disabled embryos.

Even if Deaf activists consider deafness a cultural identity rather than a disability, their arguments fall fully within the realm of the social model of disability in which the role of the impairment is fully downplayed. Disability then has only to do with disabling barriers. If we follow this line of reasoning we could also argue that to select a black or a female child is much the same thing. Still, the barriers those groups experience are totally different. Physical and cognitive impairments impose very different limitations and allow for radically different social and subjective experiences, which cannot be compared with those resulting in the same disabling barriers. But above all, a more significant factor is that impairments are not cultural traits. I do not see any problem in acknowledging that culture imbues the experience of any given impairment in one form or another. Thus impairments will be more or less disabling according to the social barriers faced by the individuals. From the point of view of the exclusion and stigmatisation there are similarities among women, black and disabled people, their different experiences notwithstanding. One must also admit that a disabled experience underlies any impairment, which imposes not only social limitations but also functional losses and limits. Unlike gender or race, impairments inflict 'inherent limitations on an individual's ability to access and make use of various opportunities' (Anstey 2002: 286). Evidently, this does not imply that limitations result only from the impairment. Rather, it implies that, unlike those constraints imposed by gender or race, disabling impairments set their own limitations which go beyond that of social barriers. Homosexuals, women, black people or other minorities can suffer from prejudice and discrimination; however, deafness and other impairments impose rather more severe restrictions on individuals. Oppression faced by disabled individuals is different and more complex than the discrimination faced by women or black people. As Shakespeare (2006) observes, there exist physiological and psychological differences between men and women, but it is difficult to argue that women are less capable owing to their biology. Gender always involves a social ascription not necessarily related to the individual intrinsic biological make-up and inherent abilities. Likewise, in the absence of discrimination, being homosexual, female or black does not put an individual at a disadvantage. This is certainly not the case with disabilities. Even with the absence of social barriers, to live with some impairment can severely limit the life of an individual. If other minority groups experience social oppression, disabled people also have to bear the intrinsic – more or less severe – limitations imposed by their impairments. They have less flexibility and fewer choices than women, homosexuals or black people. Any given ability cannot be transformed into disability, and no change in values or mentality can change disability into ability. Even if disabling barriers were eliminated the problems related to the impairment can at most be minimised but never fully eliminated.

Debates surrounding cochlear implants reproduce several issues existing in discussions around deaf embryo selection. For Deaf activists, cochlear implants represent another cultural invasion by the hearing world, a new chapter in the history of 'Deaf subjugation' and the last form of 'hearingism'. For them, cochlear implantation can be likened to a device to cure homosexuality or, similarly, for Spanish or Chinese citizens living in the US to be solely English speaking. Some go as far as considering the implant a 'cultural genocide' (quoted in Young 2002), the 'Oralism's Final Solution!' (quoted in Campbell, F. 2009: 94). It is, they argue, a 'child abuse' that 'robbed the Deaf child of their priceless gift of deafness' (ibid). They harshly oppose what they called the 'compulsory abolition of deafness' (ibid). One former president of the National Association of the Deaf even compared the practice with the Iraqi invasion of Kuwait or the beating of a blind man to make him see stars (Tucker 1997). Here, as in the case of selecting deaf embryos, one can see a prevalence of the medical model in public approaches and media portrayal of the issue, which partly justifies the anger of the Deaf community. The implant is defined as 'revolutionary', the 'miracle of life' or the 'magic technological cure' (Power 2005) that will defeat this terrible disability. Deaf self-advocates have every reason to be sceptical about the publicity surrounding the device and to debunk media portrayals and their ableist underpinnings (Blume 2010; Campbell 2005a, 2005b). Yet there are also some problems and contradictions surrounding the rejection of the implant I would like to address.

If, according to the social model, deafness is not an impairment but a cultural trait, then there is no motivation to try to avoid the impairment diminution through the implant (Shakespeare 2006). In addition, regarding the membership of Deaf culture, there is nothing to prevent a hearing child learning sign language and consequently being able to share the Deaf culture of his or her parents. Of course, Deaf activists could counter that, as one reporter remarks, in Gallaudet University (the well known US University for the Deaf) anyone who uses the device 'runs the risk of being shunned' or that 'kids who try to speak in Deaf schools are ridiculed' (quoted in Tucker 1998: 9). In my opinion, however, this argument speaks not *for* but *against* Deaf culture and the lack of tolerance frequently involved in radical identity politics. For an individual to wear a hearing device should not be derided as 'passing' or as being 'one who cozies up to the opposition' (Padden and Humphries 2010). Identity politics can be an important political strategy but it also has its drawbacks. For it hides identitarian ambivalences and promotes intra-group homogenisation. Nonetheless, not every deaf person thinks in the same way and neither should they be deemed a traitor by thinking otherwise. Some recipients of the implant reacted angrily when asked if they believed the implant 'takes away [their] Deaf pride'. One respondent asked: 'What the hell is Deaf pride? Proud not to hear your child's voice, pianos, the birds in the trees?' For him, that is not 'pride; it's bull-headedness and selfishness', and he declared to be 'proud to be part of the community as a whole, not to a "club" of narrow minded people' (quoted in Tucker 1998: 9). Judy Singer, who coined the term 'neurodiversity' within the context of autistic self-advocacy, wittily comments on the 'dark side' of identity politics due to 'its eternal victimhood, its infantilism, its demand for unconditional love and acceptance without concomitant adult self-reflection, self-criticism, a measure of stoicism, and a willingness to see light and dark in oneself as well as in "the Other"' (Singer 2007).

Additionally, I believe the most important point at stake in the cochlear implant debate is the issue of the child's interest; if a hearing child (through the use of the implant) can actually participate in Deaf culture, why are some so vigorously opposed to it? One cannot overlook the consequences of denying the implant. I agree with Savulescu (2009), who opposes the refusal of cochlear implantation, although he accepts the parent's right to select a deaf embryo. In the latter case, no child is harmed. No child is made worse off than he or she would otherwise have been. He argues that competent adults have the liberty to make controversial choices. However, when it comes to parents making choices for their children things are very different. One has always to choose what it is best for them and, as such, if an ethical position such as this is adopted, one in which the child's best interests are represented, then it is difficult to argue that deliberately ensuring that a child remains deaf is better for the child when in actual fact it narrows the scope of the child's choice and infringes her right to autonomy and to an open future (Davis 1997; Kermit 2009). Ethically, the choices of one's children later in life should be maximised[23] and, obviously, with a cochlear implant the hearing world becomes an option that adds to the Deaf world. However, without the implant, this option is restricted (Brusky 1995)[24] and the child will be worse off. Without the opportunity of hearing speech the child will not be able to participate in the dominant culture and this limits his or her options.

Along this line of reasoning, largely identified with radical versions of the social model of disability, one finds assertions such as an aeroplane is to non-flyers what a wheelchair is to non-walkers, since both are mobility aids (Oliver 1996; critically, see Shakespeare 2006). This argument is simply fallacious and has only rhetorical power, for walking is part of the normal functioning of our species, while flying is not. One cannot equate the non-flyer with the non-walker and say that it is only a question of mobility aids. The category of non-flyers encompasses all human beings, while non-walkers are individuals with some impairment that hinders them from walking. The same applies to deafness. Even a place such as Martha's Vineyard in the 19th century, an isolated community in the US where deafness was common and in fact Deaf people would not experience disabling barriers (Groce 1999), does not equalise hearing and Deaf people. Deaf people will still have an impairment and experience some form of disability when leaving their Deaf community. There is no symmetry between hearing and Deaf people. Non-disabled people can make a choice whether to use wheelchairs or learn sign language; however, this is a choice that a disabled individual does not have the privilege of making. Children with cochlear implants can choose not to use them for whatever reason (medical, political, cultural, ideological or technological) if they wish to. They can resolve to become deaf and turn off the hearing device. Deaf children without cochlear implants, on the other hand, have the right to decide for themselves whether they want to belong to one of the two cultures precluded. This creates an asymmetrical situation in which disabled children have fewer options. Therefore, I believe Savulescu (2009) has every reason to argue that, to his knowledge: 'no hearing adult has ever freely chosen to become deaf. But it would be easy to achieve. So the cochlear implant affords the deaf child an extra option: to be deaf or hearing later in life'.

Legislating deafness

Another important aspect at stake in the debate on cochlear implants is the already mentioned issue of anti-discriminatory legislation. For, as Tucker (1997) observes, in the not too distant future, courts might refuse to apply legislation to individuals who refuse to cooperate with biomedical procedures that might cure or, at least, alleviate their disability. This is a pressing argument in the debate around the implantation of hearing devices, especially because in the case of deafness, providing extensive accommodations can be difficult and expensive. According to Tucker (ibid), even by 1997 the annual cost of educating a child in a school for the deaf was estimated to be three times higher than in a self-contained class for the deaf and 10 times higher than educating him or her in a regular classroom ($35,780, $9,689 and $3,383, respectively). Tucker observes the enormous backlash of legislation such as the ADA by the business sector for which those laws provide 'special benefits' to individuals with disabilities (quoted in Tucker 1998: 12).

Such 'business-friendly judiciary' meant that the employers won almost all of the cases brought under the ADA (Bérubé 2009). In this context, it is likely that more and more courts will refuse to provide accommodation to individuals who do not take all possible steps to mitigate their disabilities (Cox 2009; Rosenthal 2002; Shaw 2002). The United States Equal

Employment Opportunity Commission (EEOC), responsible for enforcing the employment section of the ADA, establishes that the determination of disability should be made without considering the effect of medical treatment on the individual. However, this requirement is, in some cases, not adhered to. For example, in *Coghlan v H.J. Heinz Co.* (1994) the court argued that an insulin-dependent diabetic could perform major life activities and is therefore not disabled (Tucker, 1998). Also in *Pangalos v Prudential Insurance* (1997) the court denied coverage under the ADA to an individual with ulcerative colitis, arguing that it could have been cured with a colostomy, which he refused to have.

Tucker believes that this attitude may soon extend to individuals whose deafness could be alleviated with a cochlear implant, a situation which has not yet arisen. She does not discuss the fact that at the present time the implant needs to be placed in children under the age of five to maximise the chances of success. Hence, you cannot blame an adult who did not get the implant as a child (it being his or her parents' choice) for not taking all the steps necessary to mitigate their child's disability and deny the accommodation. It would be unfair to suggest that a deaf adult, in this position, falls into the category of having an 'elective disability' for, as we saw, it was not his or her choice. Whilst agreeing with Tucker (1997: 36) that it is 'difficult to balance principles of fairness and equality with economic realities' and that 'individual needs must be balanced against societal concerns' I do not agree that cover under the ADA or other anti-discrimination legislation should be denied to individuals who do not take any possible steps to mitigate their disabilities. As discussed above, this sort of reasoning can be dangerous in that it may be a trap for some people with disabilities, as *Sutton v United Airlines Inc.* (1999) shows. No adult person should be legally forced to have technological procedures and surgeries to alleviate their disability and cure the disabled body. Moreover, based on the Supreme Court's *Sutton* decision, many courts read the ADA to exclude individuals who refuse corrective surgery and other biomedical procedures. However, as Cox (2009) convincingly argued, *Sutton* suggests exactly the opposite reasoning, that disabled individuals should not be excluded from the provisions of the ADA even if any given biomedical technology might alleviate or remove their limitations.[25] Hence, the ADA requires that employers remove disabling barriers at the workplace rather than forcing individuals with disabilities to remodel themselves to adapt to existing norms or, in the words of Douglas Martin, 'we will not change to fit the mold . . . Instead we will destroy the mold and change the world to make sure there is room for everyone' (quoted in Cox 2009).

Cost-benefit arguments are frequently mobilised to defend selective abortion for disability. There are a number of unexamined presuppositions about prenatal diagnosis and selective abortion, such as the suggestion that disabled people have an inferior quality and enjoyment of life, that raising a disabled child is always a burden and that society has the right to decide who is better off not being born. However, where are we going to draw the line between the lives that are considered worth living and those that are not, and who is going to make that decision? For many people conditions such as Down's syndrome or spina bifida are intolerable and they would choose to abort a child with those conditions. They may be surprised by statements such as that made by Jason Kingsley, an adult with Down's syndrome and co-author (together with Mitchell Levitz who shares his condition) of *Count Us In*. When aged 17 he wrote about his mother's obstetrician, who had said (quoted in Bérubé 2010: 108) that he would never grow up to 'have a meaningful thought':

> He never imagined how I could write a book! I will send him a copy . . . so he'll know. I will tell him that I play the violin, that I make relationships with other people, I make oil paintings, I play the piano, I can sing, I am competing in sports, in the drama group, that I have many friends and I have a full life.

Even more dramatic would be the selective abortion for dwarfism or embryos with potential birth defects caused by thalidomide. Many individuals living happy and meaningful lives with those conditions[26] would be enraged by the idea that it might be better that a foetus be aborted because it will be born a dwarf or have deformities as a consequence of thalidomide use. But, as Saxton (2010) observes, how long will it take until selective abortion will be used to eliminate or enhance other genetically or socially idealised features such as intelligence, beauty, height, race or sexual orientation? Among the autistic community the possibility of genetic tests for autism has been received with outrage and defined as genocide (Nelson 2004). Bioethicist Arthur Caplan wrote an article in 2005 about the subject with the provocative title: 'Would you have allowed Bill Gates to be born?', alluding to the frequently observed fact that Bill Gates's behaviour exposes many traits of Asperger's syndrome and therefore he might have not been born in the event that such a test for autism had been available. Since autism is considered a spectrum, ranging from high to low functioning forms, would the test have allowed for establishing the necessary nuances clearly to define in which part of the autism spectrum a given foetus will be located, or to discriminate between high and low-functioning forms (Ortega 2009b)? The issue is a difficult one with several bioethical, political and legal consequences. I fully share Susanne Antonetta's (2005: 92) sceptical comments in her book *A Mind Apart. Travels in a Neurodiverse World* about zygote screening and genetic manipulation for children: 'Would we choose only perfect children? Perfect in the eyes of whom? Our culture?'

I agree with disability rights scholars who assert that it is of the utmost relevance to challenge limiting definitions of what should be considered 'acceptably human' (Saxton 2010). As we learn from Eva Kittay, Michael

Bérubé and others who write about the joys of living with an individual with disabilities, there is something deeply human and valuable in loving somebody with a severe disability. 'Although my daughter can never be "productive" or pay back to society anything of material value, still her contributions are great', writes philosopher Eva Kittay (2001: 576–77) emotively about life with her daughter:

> Her sweetness radiates and enriches the lives of everyone she touches, those who allow themselves to be touched by her. Without her abundant and exuberant love, the world would be a more dismal place. I am beginning to fill volumes with what I have learned from her.

Those existential 'benefits' cannot be judged in terms of productivity and economic cost-benefit calculations.

Promoting economic arguments to discuss the question of alleviating disabilities is always a tricky issue, for it is always a question of priorities. Again, I agree with Saxton (2010: 126) that 'the bottom line is that the cost-benefit argument disintegrates when the outlay of funds required to provide services for disabled persons is measured against the enormous resources expended to test for a few rare genetic disorders'. I believe autism is a good illustration of this tendency. In the context of economic recession there are huge challenges for providing therapeutic and educational services for autistic people. Against this background, the potential expansion of an autism susceptibility test may help to reconfigure risk, making the parents responsible for having kept their autistic children. While at the same time autism susceptibility genes mean biocapital, autism is turning into big business for bioengineering and pharmaceutical companies (Nadesan 2005).

For these reasons, Brusky (1995), Campbell (2005a, 2005b) and others have every reason to argue that nobody should be legally constrained to undergo surgeries or other biomedical intervention to alleviate their disability. Yet I disagree with Brusky, who believes that the decision to perform cochlear implantation on a child is only 'a preference for one lifestyle over another'. For her, courts and medical professionals would impose 'their "kindness" – their preferred lifestyles'. I believe that deafness is not just a matter of lifestyle; to be deaf is not about making a different life choice than that of a hearing person. This argument neglects the point, already discussed, that, beyond the disabling barriers imposed by the hearing world, deafness remains an impairment, a functional loss. With that in mind it is essential that we undermine the ableist underpinnings of the legal system, the negative ontology of disability and the compulsion towards biomedical intervention. This viewpoint is supported by Brusky, who asserts: 'The state must never exercise its power to reinforce prejudice and discrimination against those who are physically, culturally or otherwise different' (ibid). However, in order to do that, one should not downplay the relevance of the impairment on people's lives or to advocate the view that deafness, as opposed to hearingness, is just a matter of choosing between lifestyles. To promote this argument can be counterproductive to Deaf culture for two reasons:

First, when promoting biomedical intervention as a precondition for coverage under anti-discriminatory disability legislation such as the ADA, one can lose sight of the catastrophic consequences that surgeries and biomedical procedures can have on people's lives. Extensive publicity often surrounds medical research and popular representations of cure (for the case of deafness, see Blume 2010). However, it should not be forgotten that there is a possibility of damaging side effects as a result of these so-called 'cures'. Furthermore, one should not forget the huge economic interest and pressure of pharmaceutical and biotechnological companies involved in biotechnological and bionic developments for disabled people. Neglecting the role of the impairment upon individuals with disabilities may divert public awareness away from this important matter. Courts should be sensible to these issues when judging disability cases.

Secondly, if deafness is to be defined as a lifestyle choice and not a disability (with an underlying impairment) how is it possible to justify coverage under anti-discrimination legislation? If being deaf is tantamount to being gay or black, how is it possible to defend disability discrimination? Which is then the specificity of disabled people? As Tucker observes, 'Deaf people cannot claim to be disabled for purposes of demanding accommodations under laws such as the ADA, yet claim that deafness is not a disability and thus efforts to cure deafness should cease' (Tucker 1997: 36; see also Young 2002). Some Deaf people are all too well aware of this paradox. Padden and Humphries (2010: 396) described how they met a deaf acquaintance in San Francisco who rode the subway with disabled discount but who did not like to use the discount. For him, 'the term "disabled" described those who are blind or physically handicapped, not Deaf people'. Insisting on the cultural dimension whilst denying the impairment places Deaf people in a difficult position that hampers their political claims.

Additionally, arguments in support of the social model implied in the rejection of cochlear implantation could easily be hijacked, especially in the context of budget contention and cutting funds for the health system. The social model could be used as a vehicle by which to deny public funding for parents who want implantation for their children – and reversing Brusky's caution – to argue that hearing is a 'preferred lifestyle' over deafness. While I am not aware of such a situation in legal debates concerning deafness, in the case of autism these issues have already arisen.

Autism in court

In *Auton (Guardian ad litem of) v British Columbia (Attorney General)*[27] the judges were presented with claims made by parents' advocates that the Lovaas

Technique (a form of Applied Behaviour Analysis Intensive Behavioral Intervention [ABA/IBI]) was the only evidence-based intervention that could improve the condition of their autistic children. The trial judge held that Lovaas was a necessary intervention for children with autism. The Court of Appeal upheld the court's decision, arguing that the government had discriminated contrary to section 15 of the Canadian Charter of Rights and Freedoms. The discrimination lay in 'the failure of the health care administrators of the Province to consider the individual needs of the infant complainants by funding treatment' (para 51). This, to the appellate court, constituted 'a statement that their mental disability is less worthy of assistance than the transitory medical problems of others', thus creating a 'socially constructed handicap' that worsened the position of an already disadvantaged group (para 51). The Canadian Supreme Court overturned two lower court judgments, arguing that the provincial government did not violate the equality rights defined in section 15 of the Canadian Charter of Rights and Freedoms when it decided not to cover the cost of the therapy. The court sustained that the Canada Health Act did not grant all the required medical services, only core services, leaving non-core services to the Province's discretion.

Michelle Dawson was the only intervener named in the Supreme Court decision. Dawson, who was autistic herself, was critical of behavioural therapies such as ABA, which she regarded as unethical. In a perspective which departed from the view endorsed by parents and advocates of autistic children, she argued that the 'Court is being asked to endorse a treatment of a disability which denies the autistic population and the individuals whose interests are sought to be advanced in these proceedings, the very human dignity which the *Charter* is supposed to protect'. And she continued: '[it] makes the assumption that autistic individuals can only be fulfilled (and in fact can only be "human") if they are moulded to meet society's expectation of what is "normal"' (Dawson 2004: 40–41). She compared ABA therapies with the types of surgery that alter the appearance of children with Down's syndrome to make them undistinguishable from children without the condition and therefore more acceptable to society. Canadian society probably would not approve of such a practice and not endorse its provision as a matter of Charter rights (ibid: 41). The same applies to the use of the Lovaas technique for autistic children, argued Dawson. She went so far as to compare the situation of the autistic person with aboriginal people in Canada, and stated that treating autistic children is tantamount to 'proposing that society would profit greatly if only aboriginal people became less aboriginal or not aboriginal at all' (ibid: 56). As expected, Dawson defined (2005: 120) the Supreme Court's rejection of the petitioners' claim that intensive behavioural autism treatment must be provided as a matter of Charter right as a 'singular unprecedented victory' for autistics in particular, and for people with disabilities in general:

> In a unanimous decision written by the Chief Justice of the Highest Court of the Land, the voice of an untreated autistic was not only

> welcomed, but heard, and considered, and noted. Whatever we may have to endure in the wake of this fair and cautious decision, this singular unprecedented victory is permanent, and cannot be erased from us.

Dawson's critique of the exclusive biomedical approach to the issue of disability and the ableist underpinnings of legal theories and practices falls fully within the realm of the social model of disability. I fully agree with her when she stated that: '[t]he *Charter* does not exist to promote the view that only "normal" Canadians, or Canadians striving to be "normal", can enjoy life, health, liberty, security, and dignity' (Dawson 2004: 6). She also has every right to denounce the pitfalls of ABA/IBI therapy as being at the same time the only evidence-based treatment for autism and also constitutionally protected. She also asserted that parents could be compelled to adopt the therapy and not doing so could be seen as neglecting their children (Dawson 2004, 2005; Orsini 2009). Parents who refuse to treat their children with ABA, and autistic individuals for whom the treatment does not make a difference could be deemed to be antisocial and lacking social responsibility (Baker 2006). In addition, it is intensely debated whether ABA/IBI is actually the only evidence-based therapy for autistic children. Selection bias in the original studies, as well as their limited replicability, relativised the claim that the treatment was 'medically necessary'. It is also worth noting that the effort required in establishing one form of treatment as a cure has the effect of deflecting public attention and health resources away from other forms of treatment (Baker 2011). Hence, it is important to debunk the ableist discourse and the negative ontology of disability, as well as the publicity surrounding that particular form of therapy. In this sense, the *Auton* decision can be considered as a victory for autistic people and, by extension, for disabled people in general.

However, one should not be swept away by the euphoria to such an extent, as in the case of debates concerning cochlear implants, that social model arguments become hijacked to defend neoliberal policies used to rationalise health care and social welfare programmes. Bruno Latour, the French anthropologist, in his essay 'Why Has Critique Run out of Steam? From Matters of Fact to Matters of Concern', ponders whether 'a certain form of critical spirit has sent us down the wrong path' because, as he asks, 'what has become of critique?' (Latour 2004: 225) when its arguments are being appropriated to defend conservative and neoliberal policies that place market considerations above social and health concerns. When, for instance, to deny the threat of global warming, republican strategists think the best advice is to emphasise the 'lack of scientific certainty'? Yet, this is exactly what Latour had been doing in stressing the lack of certainty inherent in the construction of scientific facts. Hence, against constructionist arguments that denounce 'excessive confidence in ideological arguments posturing as matters of fact', writes Latour (2004: 225), the danger now will come from 'an excessive distrust of good matters of fact disguised as bad ideological biases!'

However, it is the case that, as far as the *Auton* Supreme Court decision in the autism case is concerned, the social model of disability has been incorporated into legal reasoning used to overturn two lower court judgments that granted the therapy. However, there is a danger here that one might be considered naïve. Apart from the important fact of exclusively focusing on biomedical approaches of autism and the criticism surrounding the scientific effectiveness of ABA/IBI therapy, the decision also has to be placed in the context of economic recession. Against this background, conservative policies, supported (as in this case) by constructionist arguments (the social model of disability) downsize social programmes and services and promote austerity measures substantiated by neoliberal discourses of individual responsibility and disqualification of social-welfare policies. It is not only a victory of autistic self-advocates, but it is also a victory of the neoliberal state and a defeat of the welfare state. I fully agree with Orsini (2009: 127), when he writes that

> . . . [i]n an age in which governments are looking for ways to opt out of the provision of universal programmes and services, the suggestion that ABA should not be funded because it is unproven scientifically may seem particularly disingenuous given the range of programmes and services that are financially supported by governments even though they may be lacking the sufficient evidence base.

Moreover, within the disability rights movement not everybody sees the *Auton* decision as a victory without precedent for individuals with disability. Rather, they consider that relevant international human rights law was disrespected (Malhotra and Hansen 2011). The Coalition of Accessibility Advisory Committees of Ontario (COAAC) argued that progress towards human rights for autistic people 'took a huge step back with the *Auton* decision of the Supreme Court of Canada, which essentially subordinated the rights of autistic persons to appropriate medical treatment to the policy making role of elected legislatures'.[28] Likewise, the women's Legal Education and Action Fund (LEAF) claimed that the decision is 'devastating for all disadvantaged persons' and 'has turned back the clock by decades on equality rights progress in Canada' (LEAF 2004).

Issues surrounding autism and deafness are paradigmatic illustrations of the challenges within legal discourses and practices placed by individuals framing their political and legal claims around their corporeal status. In this chapter we have seen how biological constitution, particular diseases or impairments as well as different bodily performances have entered the legal setting and become the ground for protections, rights and entitlements as well as shaping different styles of political activism within neoliberal societies. Additionally, I mobilised insights from disability studies to destabilise legal personhood. This perspective, I believe, boosts existing critiques of the abstract disembodied subject of liberal philosophy of justice and its ideals of autonomy, efficiency, productivity, rational agency and individualism. It locates the concrete, material existence of the individual and his or her interconnectedness with other bodies at the core of political and legal discourses and practices. In the remainder of the chapter I would like to make some commentaries around legal constructions of the body in order to reach some general conclusions.

Disability as a challenge to a theory of corporeal justice

Law historian Jean-Pierre Baud has observed that any consequential reflection about human corporeality has been systematically excluded within the history of Western law prior to the legal challenges introduced by contemporary biotechnologies. Western legal thinking (especially in the European continental tradition) was, for nearly two millennia, dependent on Roman law categories and in particular its fundamental distinction between persons and things. This *summa divisio* between the subject and his or her body aimed at the disembodiment of law. It furthermore allowed law scholars, lawyers and judges to apprehend the human person beyond its objectivisation and its sacralisation. Medical and philosophical objectivisations, as we saw in earlier chapters of this book, underlie Western history of anatomy and medical visualisation as well as the Cartesian notion of the body *qua* corpse or machine. Sacralisation was an integral part of the resurrected body in Pauline theology. Corporeal contingency threatened the balance and order embodied in the legal personhood of the Roman citizen. He was largely defined by his will, capable of establishing contracts and circulating patrimony (Baud 1993; Santosuosso 2001). Roman civilisation created the notion of the legal person, writes Baud (ibid: 59), to 'display man in the theatre of legal life' (*mettre en scène l'homme sur le théâtre de la vie juridique*). Legal personhood emerged in civil life to replace both body and soul, since in Christianity and Greek and Roman philosophical traditions the individual was apprehended as a body housing a soul. Hence, the Roman notion of *persona* is a legal creation identified with the legal subject (*sujet de droit*) whose goal is to 'abstract' – in the etymological sense of putting outside the realm of sense reality – the human person identified with his or her body.

In short, since it is not the aim of this chapter to write a history of the body in Western law, the individual, for Western traditions of law, is not an entity of flesh and bones but a reasonable being who also possesses a body. To give an example from the history of Continental law, the Napoleonic Code, the French civil code from 1804, underlies a dualistic anthropology with influences from ancient philosophical thinking, Canon law and Cartesian philosophy. Man appears as pure will without a body, needs or desires, a strong will oriented towards freedom. Within this male and disembodied notion of legal personhood, women, children, homosexuals and sick people underwent a legal devaluation owing to their dependence on bodily passions and their subjection to various guardianship regimes. The latter were implemented to correct the deviance resulting from their corporeality (Borrillo 1994). At the end of the 19th century the disembodiment of law became the greatest achievement of French civil law. Against this background a French jurist could confidently write in an encyclopaedia article about 'person' that in a society which abolished slavery 'the physical man has only a place in natural history or in physiology', not in law (quoted in Baud 1993: 104).[29]

In the second part of the 20th century ongoing debates around assisted reproduction, commodification of body parts – blood and tissues; organ transplantation – human experimentation and genetic engineering have transcended the traditional Roman and civil law distinction between persons and things. As a consequence, human corporeality was placed at the centre of legal debates. I agree with Borrillo (1994) that many incongruences, paradoxes and misunderstandings in dealing with these topics derive from the lack of tradition of philosophical reflection about the legal status of the human body. Within a critical legal studies framework Alan Hyde (1997: 4) argues in a similar vein when he observes that there is no consensus about the body in contemporary American law. Body means an 'inconsistent and incoherent assortment of representations and visualizations, deployed to solve political problems internal to legal discourse'. Another law scholar defined US laws governing the body as 'chaos' (quoted in Curry 2002: 23). Specifically in the case of bodily autonomy there are several inconsistencies in the legal treatment of the issue that can be derived from different and contradictory legal doctrines.

Legal discourses construct the body as a thing, separated from the person. The legal subject is distinct from the body, which enables market transactions and its public regulation. According to Hyde (1997) the first legal construction of the body in the US in the 20th century corresponds to the machine body. The body is a machine owned by its owner or buyer and used by the first to raise profit. The machine metaphor has a long history within Western philosophical and also medical thought. The use of the analogy in legal discourses fits well within a notion of the body as property or a commodity. Its appraisal in terms of money evidences its commodified nature. The machine analogy was replaced by the legal construction of the body as property or a commodity, which incorporated many of the properties of the earlier model, such as fungibility, estrangement and desentimentalisation (Hyde 1997; Curry 2002). Yet different constructions of the body in legal discourses and practices have something in common. They all 'emphasize its thingness, its distance from us, that treat the body as object, property, machine' (Hyde 1997: 260). Therefore it is necessary to undermine the legal arguments that construct the body as property, machine or as a private interest or inviolable.

The legal body corresponds in our time to the commodified body of consumer culture, in which any corporeal transformation, display or pain is subsumed under the form of the contract. To find alternatives to the body as property, machine or privacy right involves, according to Hyde (1997: 6), treating people as 'embodied, that do not shy away from pain, sex, or other embodied experiences', and substituting the metaphors of machine, property or privacy right with a 'language of bodily presence or embrace'. Herring and Chau (2007) have accordingly argued that human bodies exist in a state of dependency and interconnectedness with other bodies from the moment of intrauterine existence to death; they are in perpetual interaction with the surrounding environment and are not immutable entities but are in constant transformation. Scholars such as Shildrick (2010) have opposed the idea of corporeality as a standardised and self-contained entity expressed in the ethico-legal notion of 'bodily integrity'. Experiences provided by prosthetics, organ transplantation and other forms of body modification challenge notions of the body's 'natural' constitution and limits, as well as the normative structure of human embodiment presupposed in property law. The body is 'never self-complete and bounded against otherness, but is irreducibly caught up in a web of constitutive connections that disturb the very idea of human being' (Shildrick 2010: 13). These are insights also present in the phenomenology of the body and other philosophical and psychological theories discussed elsewhere in this book. The novelty here is to apply these insights to deconstruct the legal status of the body based on property, rights to privacy or integrity that we find in debates concerning the property of removed body products, conceptions of genetic privacy, the Human Tissue Act and organ donation.

Legal texts' assumptions about the body overlook the interdependence and interconnectedness of bodies (intercorporeality), which prevents the identification of them, in a straightforward sense, as 'ours'. Hence, the language of property or ownership is confusing and above all morally unproductive. It is a futile exercise to identify the relationship we have with our bodies as equivalent to the one we have with material possessions. The legal procedures when dealing with the former cannot be analogous to those procedures found when dealing with property trespassing or possession theft. Therefore, I fully endorse Alastair Campbell's (2009: 15) critique of legal accounts of the body, which circumscribe it within the rights and obligations derived from property law:

> It seems unlikely that an assertion of property rights over our body, of the kind we hold over (say) a tract of land or a library of books, is the most effective way of defending ourselves against unlawful incursions into our personal space.

Legal narratives and their 'corporeal' metaphors – the body as machine, property, commodity or privacy right – reinforce the traditional mind/body dualism predominant in Western philosophical thinking and the history of medical visualisation. Objectivisation and commodification of the body and the resultant devaluation of physical experience derive from the property relationship in liberal legalism (Halewood 1996). As noted earlier, liberal constructions of legal subjectivity treat the body as an object divorced from the subject and not as integral to personhood. This separation deprives subjects from all particularities and marks of embodiment necessary for establishing the formal equality of legal subjects. In our legal system the rational and volitional elements dominate the corporeal and emotional and obedience to the law derives from the 'evidentness or self-presence of legal norms to the legal subject's power of reason' (Cheah and Grosz 1996: 3). Legal practices which construct the body as property promote a notion of corporeality as privately alienable by markets rather than a conception which privileges embodiment and the subject's physical experiences. Law should instead look for theoretical models that display and endorse the interconnectedness and interdependence of bodies. Taking embodiment as a starting point implies shifting the liberal subject/object model of property and looking for alternative types of corporeality that are not based on property relations. Within a 'theory of corporeal justice' (ibid: 25) differences in bodily make-up of subjects before the law should be respected and legal narratives based on our embodied subjectivity should replace civil and property law accounts of corporeality.

It is beyond the scope of this chapter to offer a fully developed 'theory of corporeal justice', as impressionistically suggested by Cheah and Grosz (1996). However, this proposed path is designed to address several important challenges for legal theory and practice presented by somatic culture and bioidentities. I have analysed relevant political and legal implications that arise from citizens' framing of their political claims around their corporeal status. One can ask how biological constitution, different bodily performances, diseases/impairments or genetic make-up can become the basis for rights, entitlements and protection claims, as well as constituting a major vehicle for political mobilisation and activism in advanced liberal democracies. There is obviously no straightforward answer to this question, since it involves taking into account different forms of health-related activism and identity politics, which differ according to social and political contexts, economic status, types of disease or impairment and styles of scientific thinking.

In this chapter I have explored some of the intersections between our embodied subjectivities and notions of citizenship. Bacchi and Beasley (2002) have developed the idea of 'embodied subjectivity' as a trope through which to analyse these points of intersection. I, for my part, have addressed this relationship using the notion of 'biological citizenship', originally thematised by Adriana Petryna and developed by Nikolas Rose and other scholars (Petryna 2002; Rose 2007; Rose and Novas 2005). Petryna's seminal analysis of biological citizenship convincingly showed the limits of the idea of the total availability of the body, when it is used to bargain for citizenship and other compensatory claims. Whilst in the post-Soviet society biological citizenship largely embodies the failure of political and social citizenship, in advanced liberal democracies it designates a form of consumer citizenship. In both contexts, however, biological citizenship constitutes a departure from traditional Marshallian notions of citizenship but also, especially in the former context, a devaluation of citizenship. In contemporary neoliberal societies biological citizenship can coexist more or less harmoniously with traditional forms of citizenship. However, it is the only form of citizenship granted to several groups excluded from citizenship owing to their nationality, ethnic affiliations, socio-economic status or bodily make-up. In this sense it represents a perilous alternative to political citizenship.

An instrumental understanding of corporeality, expressed in the notions of body as machine, property, commodity or private right, informs our constructions of citizenship and also our legal practices. Against this background individuals are presumed to have control over their bodies as autonomous citizens, while those who do not have this power are considered to be lesser citizens. I focused on the notion of dis-citizenship to appraise this situation of unequal citizenship, such as that experienced by individuals with disabilities and other excluded groups and minorities. Disabled citizenship or dis-citizenship offers the most radical challenge to embodied citizenship as individuals with disabilities cannot for the most part match the ideals of efficiency and productivity demanded by active consumer citizenship.

Several scholars within critical legal studies, critical race theory and feminist legal theories have criticised the subject of liberal philosophy of justice as disembodied, genderless and raceless, and demanded a more embodied, gendered and contextualised account of justice that undermines liberal legalism and abstract notions of personhood. In this chapter I have shown how the disability studies approach can usefully undermine the ideals of autonomy and self-sufficiency present in contemporary legal constructions of personhood. I believe that the challenge imposed by those accounts can radicalise critical legal studies' critique of legal personhood. As we saw, disability destabilises the disembodied subject of liberal political and legal theory by subverting the notions of efficiency, productivity, rational agency, autonomy and individualism, which underpin liberal understandings of justice and citizenship. It does so by placing the concrete, lived, fleshy existence of the individuals and their embodied experiences at the centre of political and legal theories and practices, thereby dismantling normalising ideas of corporeality and ableism.

Informed by the analytical tools of critical legal studies, critical race studies and feminist legal theories, scholars working with disability legal studies and critical disability theory approaches have gone beyond merely analysing court decisions or legal policies to the actual scrutinising of the legal construction of disability and ableism. CLS' traditional critique of rights denounced that all claims to rights would be acknowledged to the extent that they maintain the political integrity of the body politic. Any apparent victory in the struggle over civil rights bolsters mainstream legal and political theory (Cheah, Fraser and Grbich 1996). Critical race theory scholars have drawn on the notion of 'interest convergence' to stress the way white people tolerate black empowerment only insofar as it aligns with white interests (Bell 1980; Milner 2008). Hence, when black people's interests are in opposition with those of white people, it becomes more difficult to pursue racial equality and fight racism (Leigh 2003). People in power might sustain anti-discrimination policies and practices and provide certain accommodations as long as they do not alter their way of life and the socio-economic and political privileges they enjoy. Therefore, interest convergence, write Castagno and Lee (2007: 10), 'exposes the selfishness behind many policies and practices that may advance greater equality'.

Scholars working with a critical disability studies approach have incorporated this CLS and CRT critique, and the notion of interest convergence to explore the limits of liberal tolerance to disability. This attitude is obvious for instance in the relationship between paternalist attitudes, the ideology of ableism and the subordination of disabled people. Sympathetic attitudes towards disabled people coexist with their social and economic subordination and their depiction as dependent, asexual, helpless and economically unproductive (Campbell, F. 2009). Within this ableist context, impairment is seen as the only source of the disability problem downplaying the socio-political responses to it and society's disabling barriers. Against a socio-economic background where the support and financial aids provided by government welfare systems to facilitate full equalisation of opportunities are becoming more and more difficult to obtain, disabled people have to appeal to the court system for remedies and services to attenuate the effect of the impairment and to secure monetary compensation for their loss. As noted above in the case of race, disability benefits and compensations are granted insofar as they maintain and strengthen the ableist normalisation of bodies. The power of law to generate disablement and to 'legalize ableism' (Mor 2006) demands that the disabled litigant act according to the norms of ableness. This implies that the disabled plaintiff performs or acts out his or her disabled role; that is, that he or she reduces his or her social and subjective experience of living with a disability to the limitations imposed by the impairment. Legal renderings of disability require the disabled to perform the impairment, to adopt the deficiency in the body and to embrace the narrative of suffering. Any alternative readings of living with impairments which denote joy or ambivalence are not readily accepted by the law. Legal constraints 'make it impossible for [those] stories to be heard and recognised' (Rovner, quoted in Campbell, F. 2009: 132). Let us illustrate this situation with a practical example from an interesting area of civil law, the law of torts and, more specifically, wrongful birth/life actions.

Briefly, wrongful birth actions involve situations where parents of impaired children sue for wrongful birth on the basis of misdiagnosis or failure to detect genetic defects; whereas wrongful life involves actions brought by people with impairments on the basis that they were born with disabilities owing to the actions or omissions of a doctor or another party (Campbell, F. 2009: 148). Examples of the latter within US and Australian courts frequently involve cases of women contracting rubella or other diseases during pregnancy for which they had not been tested or screened and as a consequence gave birth to children with intellectual and/or physical disabilities. Campbell made an important point when analysing those court decisions. The problem is not only the need to seek compensation and financial support for disabled people and their families – which in itself attests to governmental failure in providing assistance to those in need – but that they have almost no option but to seek this compensation through a legal process. What is problematic – and disabled people pay a toll for that – is that claimants and their legal supporters are thus obliged to operate with negative ontologies of disability. Disability is presented here as a personal tragedy and disablement is reframed in terms of mitigation as a means to contend wrongdoing and claim compensation. Rights and compensations are granted to the extent that they maintain and reinforce ableist normativity.

The first traditional wrongful life suit was *Gleitman v Cosgrove*, brought before the New Jersey Supreme Court in 1967. The plaintiff unsuccessfully claimed physician negligence in not informing her of potential harm to her foetus after she developed rubella in the second month of her pregnancy (*Gleitman* 1967: 690–91). The court denied the claim on the grounds that the harm done was incalculable, citing 'countervailing public policy supporting the preciousness of human life' (Burns, quoted in Carey 2005: 2). Legal reasoning generally assumes that every child has the right to be born as a whole functioning human being; a right denied in this case owing to the doctor's negligence. The argument was not that the child should have been born without defects, but that 'he would not have been born to suffer with an impaired body', following the child's later argument that he 'would be better off not to have been born' (*Gleitman* 1967: 692). In order to obtain compensation the families and the legal team in such cases have to attest to the intolerability of the impairment, demonstrating why living with it is inherently harmful and, consequently, why it might be better not to have been born. Hence, compensation and aids are granted only insofar as they reinforce ableist prejudices and the negative ontology of disability. In fact, as Justice Timothy Studdert of the New South Wales Supreme Court determined when judging on the wrongful life cause of action brought in *Edwards v Blomeley* (2002), *Harriton v Stephens* (2002) and *Waller v James* (2002),[30] to grant damages based on wrongful life would harm the self-esteem of disabled people. For him, 'There are many disabled members of society who lead valuable and fulfilling lives . . . to all such persons the notion that non-existence may be considered preferable to living with disabilities would surely be perceived to be offensive' (quoted in Campbell, F. 2009: 155). Campbell is completely right when she states that, notwithstanding the positive or negative outcome of wrongful birth and wrongful life judgments, legal rendering of disability always draws on expressions such as 'bearing the burden of affliction', 'suffering' and an 'arduous responsibility', thus replicating and reinforcing the idea of disability or impairment as intolerable and inherently negative (ibid: 157).

Wrongful birth and wrongful life actions offer an illustrative example of the production and perpetuation of disablement within the legal setting. The apparently positive outcome for individuals with disabilities and their families comes at a high price for them, for they are obliged to wear the 'clothes of a disabled identity' (Holloway, quoted in ibid: 35). This means acting according to an ableist normativity, performing the disabled role as inherently negative and acknowledging that an impaired life has less worth than a 'normal' life. What appear to be the positive outcomes of successful wrongful birth and wrongful life litigation are, in my view, thus often a trap. In this sense, one should go beyond the presentation and evaluation of court decisions and legal policies, regarding the rights of individuals with disabilities, as simply individual remedies, and to explore instead, as CLS and CRT critics have insisted, how struggles and resistances are domesticated within the legal setting. Claims to justice as rights to be recognised or settled in legal disputes have alienating and disempowering effects for marginalised people and especially for individuals with disabilities. Moreover, they attest to the failure of governments and the welfare system to provide support and financial aids for disabled people, leaving them with no option but to seek remedies and services to attenuate the effect of the impairment and monetary compensation within the uncertain and generally expensive setting of the courts (Campbell, F. 2009). Finally, and even more problematically, they obscure the fact that disability benefits and compensations are only settled insofar as they not only maintain but also reinforce the ableist normalisation of bodies, prolonging the stigmatisation and exclusion of individuals with disabilities from full equalisation and participation in mainstream society.

My interest in this chapter was to trace instances in which the corporeal existence of individuals is incorporated into the legal process, including the construction of legal narratives replete with their corporeal metaphors of machine, property, commodity and privacy rights. These narratives are informed by the dominance of property relationships within liberal legalism resulting in the disembodiment of legal personhood. Liberal political and legal philosophy regards the body as separated from the subject and not relevant for describing personhood, thus prolonging the traditional mind/ body dualism prevalent in Western philosophical and medical thought within legal theory and practice. Against this socio-legal background, the relationship with one's own body becomes commensurate with the relationships we entertain with material possessions as regulated through property rights and law. The last consequence would be the total commodification of human personhood, including our bodies, promoted by a free market philosophy (Campbell, A. 2009). These legal constructions of the body, which disregard the individual's bodily nature and subjective embodied experiences, including the fact that bodies are dependent and interconnected with other bodies and with the environment, are disempowering and alienating. Theoretical models which give pre-eminence to the embodied nature of the legal subject shift the property paradigm and its dualist metaphors and fully acknowledge and respect the corporeal differences of the subjects before the law.

I believe that disability studies can offer an important contribution to the development of such theory of corporeal justice and radicalise the critique of the abstract disembodied personhood of liberal philosophy of justice undertaken by CLS, CRT and feminist legal studies. Disability undermines the ideals of autonomy, efficiency, productivity, rational agency and individualism implicit in liberal notions of justice and citizenship. Those principles of liberty and autonomy do not reflect the actual lived experience of people living with disabilities. Independence and autonomy are projections of an ableist society. The illusion of the autonomous, independent individual has been destabilised by disability scholars (who draw on insights developed by feminist literature). By exposing the interconnectedness of people and their bodies they challenge the claim of autonomy understood as independence and sovereignty and move towards notions of interdependence and relationality. Dependence is a liberal fiction that hides a whole range of support and dependency systems; for we are all dependent, interconnected, relational beings (Carey 2009). In this sense disability studies have taken the farthest-reaching step in theorising these issues. In post-modern and post-structuralist academic circles, Gilles Deleuze's claim of the 'indignity of speaking for others'[31] remains an alluring ideal of emancipatory thinking and a critique of paternalistic and patronising attitudes (Bérubé 2010: 103). I fully agree with Bérubé that, for some people with severe cognitive disabilities, the surest form of recognising their rights and dignity is precisely by assuming their deep dependency and delegating to their guardians the full rights and responsibilities of speaking for another. Strong emphasis on surrogacy and an ethics of guardianship here replaces the ideals of autonomy and self-representation as they become the surest means of granting people with disabilities not only social and economic entitlements as well as equal opportunities in education, but also political and civil rights on the basis of full equality.[32]

Notes

1 The body between constructionism and phenomenology

1 I am thinking of authors such as phenomenologists Edmund Husserl, Maurice Merleau-Ponty, Erwin Strauss, Samuel Todes and Drew Leder; the pragmatism of William James and John Dewey; Donald W. Winnicott's psychoanalytic theory; and the ecological psychology proposed by James Gibson and Edward Reed (see Costa 2004; Bermudez 1998; Butterworth 1998; Campbell 1998; MacLeod and Pick Jr 1974; Husserl 1952; Merleau-Ponty 2002; Dewey 1929; James 1890, 1904; Leder 1990; Shusterman 2012; Strauss 1935; Todes 2001; Winnicott 1990, 1992, 1999).

2 See Depraz (1997) on the difficulty of translating the German notion of *Leib*.

3 I therefore disagree with authors such as McLaren (2002), who identifies an explicit rejection of mind/body dualism in Foucault's work.

4 It could be argued that one thing is what the Greeks thought about the self (in dualistic terms); another is what Foucault thinks about the self. He might well refer to the Greek self without adopting its ontological foundations. However, Foucault's analyses of the Greek self are not only historical, but also (and, in fact, primarily) philosophical. Moreover, as he states in his Collège de France lectures on *The Hermeneutics of the Subject*, the ancient ethics of the self may represent an alternative to neoliberal forms of selfhood (Foucault 2005: 252). If this is the case, then the dualist underpinnings of the Greek self are epistemological and ethically relevant when thinking about an alternative to the neoliberal self.

5 Levin (2008: 193) sees no contradiction in Foucault's account of corporeality. She argues that being disciplined and a free agent are not mutually exclusive, and therefore that the early Foucault does not preclude the later one. However, while the same individual may be a free subject and a disciplined subject in different social and historical contexts, human corporeality in both situations is the same. Therefore, the body that underlies discipline must be the same as that which undertakes techniques of the self. And the latter requires going beyond the somatic singularities of disciplines. Bryan Turner's remark (2012b: 68–69) that Foucault's exploration of the care of the self complicates all inquiries into governmentality does not solve the problem either. He recognises that the care of the body/self requires going beyond the docile bodies of disciplines, but does not perceive that, in order to do so, one would need the phenomenological body which Foucault is so reluctant to accept.

6 On the history of the concepts of action and reaction and the primacy of 'reaction' from physics and chemistry to psychiatry and politics, see Jean Starobinski's remarkable *Action and Reaction. The Life and Adventures of a Couple* (Starobinski 2003).
7 This does not mean that there can only be a phenomenology of the acting body. There are phenomenological explorations of the passive body – sleeping, prenatal or even dead body. These passive bodies represent aspects of depth disappearance, withdrawal from the egoic self and sensorimotor engagement (Leder 1990: 144–45). For a phenomenology of the corpse from a Heideggerian perspective, see Dalton (2012).
8 For a good overview of Cronenberg's films with many images see http://www.davidcronenberg.de.
9 See Sterlac's observation about the differences between his position and Orlan's understanding of the obsolete body (Zylinska and Hall 2002: 121–22). According to the Australian artist, Orlan is still caught in a form of Cartesian metaphysics. Sterlac's body obsolescence, in contrast, does not discard the body, but implies redesigning it in a technological environment.

2 The transparent body

1 In *Ideas II* (Husserl 1952) phenomenologist Edmund Husserl referred to the body as the 'zero point [*Nullpunkt*] of all perception' (see also Duranti 2010; Holenstein 1999; Petit 2010; Waldenfels 2000; Welton 1999). Most authors who hold positions close to the phenomenology of corporeality converge on this point.
2 Leder is here assuming a state of health. As the French surgeon René Leriche put it: 'health . . . is life lived in the silence of the organs' (quoted in Canguilhem 1991: 91). Pain, for example, interrupts the state of depth disappearance characteristic of the inner body; see Leder (1990: 70–79).
3 Leder would not deny that when we see the interior of our body, we react subjectively (with pleasure, horror, repulsion etc). What he really means is that we do not experience its functioning. I can visualise my colon or my lungs, and yet, as he writes (1990: 44), their 'processes still elude experience from within'.
4 See, among others, Annoni and Barras (1993); Ellis (2001); King and Meehan (1973); Mandressi (2003); Porter (2004).
5 See Carlino (1999); Ferrari (1987); French (1994); Grmek and Bernabeo (1997); Mandressi (2003); Park (1994, 2010); Porter (1999, 2004).
6 We should remember Descartes' fascination for anatomy and its role in the development of the body-as-corpse model, which has not been sufficiently examined (Cottingham 1986; Rodis-Lewis 1995). According to an anecdote, in response to a question about his readings, the French philosopher responded, showing a dissected calf: '*Voilà ma bibliothèque*' (quoted in Le Breton 2003: 62). Descartes' correspondence shows him to be intensely involved with anatomy during the decade in which he elaborated his philosophy. He even told Mersenne that one winter in Amsterdam, he visited butchers' shops daily looking for carcasses for his anatomical studies. See the letters of 20 February and 13 November 1639, in Descartes (1639).
7 For exceptions see Danius (2000) and van Dijck's (2005) chapter 'X-Ray Vision in Thomas Mann's *The Magic Mountain*'.
8 Figures taken from http://www.bodyworlds.com/en/exhibitions/unparalleled_success.html. The historical and socio-anthropological literature on Bodyworlds

is enormous. See, among others, Bleeker (2008); Connor (2007); Durbach (2012); Hirschauer (2006); Jespersen, Rodriguez and Starr (2009); Lantos (2011); Linke (2005); Stern (2006); van Dijck (2001, 2005).

9 In an attempt to find out what Bodyworlds visitors really think, independent visitor polls were carried out in different exhibition sites, from Japan to London and Los Angeles. Eighty-five per cent of the visitors reported that 'they knew more about the human body after their tour through the Exhibition'; and 54 per cent stated that it 'made them think more about life and death'. Finally, 79 per cent felt 'deep reverence for the marvel of the human body'. See 'Visitors' Reactions to Bodyworlds', available at http://www.californiasciencecenter.org/GenInfo/MediaRoom/PressReleases/BWmedia/PressReleases/VisitorSurvey.pdf.

10 There is also extensive historical and socio-anthropological literature about the Visible Human Project. See, among others, Bleeker (2008); Csordas (2000); Hirschhauer (2006); Lizama (2009); Stern (2006); Treichler, Cartwright and Penley (1998); van Dijck (2000, 2005); Waldby (2000).

11 The list is huge; the following are only some titles: *Fragments for a History of the Human Body* (4 volumes); *The Fragment. An Incomplete History; The Body in Parts: Fantasies of Corporeality in Early Modern Europe; Fragmentation and Redemption: Essays on Gender and the Human Body in Medieval Religion; Der versehrte Körper: Revisionen des klassizistischen Schönheitsideals; Körperteile: Eine kulturelle Anatomie*; and, in Brazil: *O corpo impossível: A decomposição da figura humana de Lautréamont a Bataille.*

12 The literature on the topic is very extensive. See, among others, Cooper (2008); Kimbrell (1993); Sharp (2006); Sunder Rajan (2006); Waldby and Mitchell (2006).

3 From ascesis to bioascesis

1 Historian of religion Vincent L. Wimbush concludes that 'no one text, no one historical figure or group from antiquity, and no particular type of practice could adequately define or typify asceticism. The interplay of practice and motive and the seemingly infinite number of combinations and degrees of tensions in the dynamic between practice and motive in different settings were found to be far too complex to allow any of us to be comfortable with generalizing the phenomenon from one focus or area of research' (1990: 1).

2 In the specific case of Christian late antiquity, Valantasis (2008: 44–47) distinguishes five types of ascetic subject: the combative subject, the integrative model, the educative model, the pilgrim and the revelatory subject.

3 For a historical genealogy of food dietetics using the Foucauldian framework, see Coveney (2006).

4 In the Weberian interpretation, early Christian asceticism constitutes a set of other-worldly practices, indifferent to human affairs, and will only attain an inner-worldly dimension with ascetic Protestantism (Schluchter 1988; Weber 1905).

5 Nikolas Rose uses the term 'advanced liberalism' and 'advanced liberal' democracies in connection with specific technologies of government characteristic of the last third of the 20th century in Europe and the US. These technologies received support across the political spectrum, and concerned a range of domains, from crime to health, in which individuals became governed 'at a distance' and

through their own freedom. 'Advanced liberal rule', writes Rose, 'seeks to degovernmentalize the State and to de-statize practices of government, to detach the substantive authority of expertise from the apparatuses of political rule, relocating experts within a market governed by the rationalities of competition, accountability and consumer demand. It does not seek to govern through "society", but through the regulated choices of individual citizens, now construed as subjects of choices and aspirations to self-actualization and self-fulfilment' (Rose 1996: 41; see also Miller and Rose 2008).

6 In the lectures on *The Hermeneutics of the Subject* Foucault puts it in a less straightforward way: 'Throughout the long summer of the Hellenistic and Roman thought, the exhortation to care for oneself became so widespread that it became, I think, a truly general cultural phenomenon' – which was at the same time an 'event in thought' (2005: 9). Critics have accused Foucault of providing a 'fatally linear' history (Porter 2005: 126). Boyle (2012), on the contrary, has convincingly argued that the care of the self is a general normative practice that takes different forms in different historical periods.

7 See for instance the two collective volumes edited by Taylor and Vintges (2004), and Binkley and Capetillo-Ponce (2010).

8 The interpretations that draw on the terminology of technologies of the self to discuss neoliberal identity formations generally refrain from judging the increasing interdependence of 'somatic ethics' and the 'spirit of biocapital', and limit themselves to tracing the 'emergent forms of life' and the 'possible futures' that result from such interdependence (Rose 2007: 258; critically, see Cooter and Stein 2010; Slaby and Choudhury 2012).

9 The literature around different aspects of governmentality within neoliberal societies has grown steadily for the past 20 years. See for example Burchell, Gordon and Miller (1991); Miller and Rose (2008); Rose (1999); Barry, Osborne and Rose (1996); Bröckling, Krasmann and Lemke (2011); Nadesan (2008).

10 For a good overview see Rothblum and Solovay (2009).

11 The following paragraphs are adapted from Ortega (2011).

12 *Brain Gym for Business*; *Hands On. How to Use Brain Gym in the Classroom*; *Cross-train your Brain*; *Build Your Brain Power*; *Brain Power. A Neurosurgeon's Complete Program to Maintain and Enhance Brain Fitness Throughout Your Life; Brain Fitness. Anti-aging Strategies for Achieving Super Mind Power* (Dennison and Dennison 1989, 1994; Cohen and Goldsmith 2002; Eiffert 1999; Winter and Winter 1987; Mark and Mark 1991; Goldman, Klatz and Berger 1991).

13 Moreover, Rose has argued that reflexivity is not a feature of late modernity, but defines liberal political rationalities since the emergence of liberalism (Miller and Rose 2008: 206).

14 See, among others, Crawford (2004, 2006); Crossley (2004); Edgley (2006); Le Breton (1999); Sant'Anna (2001); Morris (2000); Lupton (1994); Glassner (1989); Hesse-Biber (1997); Negrin (2008); Gori and Del Volgo (2009).

15 In the chapter on 'The Self' of his *Principles of Psychology* (1890), William James too postulated a notion of extended self that encompassed the individual's goods, clothes and properties. James nonetheless did not believe that these external marks of selfhood were its most representative parts. In the same way, for James and several phenomenologists of corporeality, from Husserl and Merleau-Ponty to Shaun Gallagher and Dan Zahavi, the self is basically corporeal,

but again that does not mean that external appearance constitutes the individual's most authentic identity.

16 There is ample literature on the biomedical reduction of the lived body to the physical body; see for instance Young (1997), Toombs (2001), Leder (1990), Jackson (2001), Lyon and Barbalet (2001), Duden (1997), Andrieu (2002a, 2002b, 2003) and Csordas (2001).

4 Bodies on trial

1 The literature around biosociality is constantly growing since Rabinow first developed the concept in the early 1990s. For a good collection of articles on the subject see the book edited by Sahra Gibbon and Carlos Novas (2008), which includes an afterword by Rabinow. I agree with Scott Vrecko (2011), who has argued that the concept of biosociality is generally used to refer to patient advocacy, identity and activism, neglecting other social transformations arising from the new biosciences. In the previous chapter I drew on the concept of biosociality without necessarily linking it to patients' activism and advocacy. However, this link is relevant for the objectives of this chapter, especially concerning its legal implications. One could also argue that the concept has been mainly restricted to genetics, whilst the new brain sciences are important instances of the emergence of biosocialities suggested by the outgrowth of concepts such as 'neurochemical self', 'cerebral subject' and 'neurocultures'. In this chapter I take up cases of biosociality not directly connected to genetics, such as the autism rights movement or the Deaf movement.

2 For an excellent review of current debates around biological citizenship and governmentality, see Wehling (2011).

3 Their description is indeed programmatic and only partially decontextualised. They do not undertake particular ethnographies of biological citizenship and limit their discussion largely to internet sites. However, biological citizenship in advanced liberal democracies articulates mostly through the internet. In the case of autism self-advocacy the internet has been the basic form of mobilisation, since many people diagnosed with autism experience difficulties in face-to-face contact. The internet is for them akin to 'Braille for the blind' (Blume 1997).

4 'Ableism' (or the 'able-bodied') describes the structure of power that, as in the cases of sexism and racism, renders one group of people (persons with disabilities) inferior to and dominated by another group (the non-disabled people). The ideology of ableism induces social stigma and oppression against the physically and cognitively different (Campbell, F. 2009; Linton 2010; Mor 2006).

5 One of the most influential discussions is Martha Nussbaum's well known analysis of disability and justice in *Frontiers of Justice* (2006). Nussbaum develops the capabilities approach as a non-contractarian account of care, which may incorporate disability in a theory of justice. See also Nussbaum (2010); on discussions of Nussbaum's approach, see Bérubé (2010); Denier (2007); Richardson (2006); Stark (2010); Wong (2010); Young (2007). The literature on the topic of disabilities and justice is enormous, both at the theoretical and the practical level, with many ramifications and concrete issues and public policy applications. It is not the goal of this chapter to go deeper into these discussions. See among others, Kittay (1998, 2001); Silvers, Wasserman and Mahowald (1998);

Brighouse (2001); Engel and Munger (2003); Pothier and Devlin (2006); Turner (2006). Further in this chapter I will address some concrete instances of disabilities in legal contexts as an illustration of how the issue of disabled corporeality can destabilise basic assumptions about legal subjectivity.

6 Eva Kittay (1998), among other scholars within disability studies who developed a feminist ethics of care, deserves the credit for the best development of this insight.

7 The label 'Critical disability theory' has also been used to define a Foucauldian approach to disability theory and activism. The latter intends to be an alternative to the juridical notion of disability currently represented in the social model of disability. According to the social model, social and economic marginalisation experienced by people with disabilities is primarily due to structural barriers in society. This model analyses how economic relations, institutional practices and settings and cultural understandings of disability contribute to the disablement. While British disability theorists stressed class politics and the tyranny of the welfare state, their American colleagues drew heavily on the civil rights paradigm (Malhotra 2006). In this context claims to entitlement advanced by disability scholars and activists against biopolitical and governmental practices are themselves productive effects of liberal governmentality and biopower. See Tremain (2005).

8 In Canada's Supreme Court Decision in *Granovski v Canada (Minister of Employment and Immigration)* [2000] 1 SCR703 the claimant suffered from a temporary disability that prevented him from working and contributing to the Canadian Pension Plan (CPP). By the time his impairment became permanent he was excluded from his relevant benefits because of the lack of contributions following his disability. Among the issues raised by the appellant was whether the CPP infringed section 15(1) of the Canadian Charter of Rights and Freedoms because the contributions requirement acknowledges persons with permanent disabilities to take advantage of drop-out provisions in the legislation, whereby periods of disability are not counted when calculating the contribution. Yet it fails to take into account the fact that persons with temporary disabilities may not be able to make contributions for the minimum qualifying period because they are physically unable to work. The appellant was unsuccessful both before the Pension Appeals Board and before the Federal Court of Appeal. Writing for the court, Justice Binnie uses a formal equality model that compares the claimant to the group of the permanently disabled, whose benefits he claims to have the right to share, although he does not fit into that category. She thereby dismisses the claimant's argument about his temporary disability having the same effect on employability as a permanent disability and therefore having the right to the same benefits. He was not sufficiently disabled to take advantage of a statutory regime directed toward disability. Using equality as a comparative concept reinforces the medical model of disability that does not take into account the claimant's assessment of the reality of his or her situation. Thus, the comparison 'between disabilities, or between the disabled and the abled', write Gilbert and Majury (2006: 290), 'is not only an impossible task but one that perpetuates dominant notions of normalcy'. See also Pothier (2006).

9 The decision of the Supreme Court of Canada in *Eldridge v British Columbia (Attorney General)* [1997] 2 SCR 624 involves the situation of three deaf people

who were not provided with an interpreter for health care services in British Columbia. The appellants contended that the absence of interpreters impaired their ability to communicate with their doctors and other health care providers and thus increased the risk of misdiagnosis and ineffective treatment. On 9 October 1997, the Supreme Court unanimously decided that the Government of Canada should pay for sign language interpreters when Deaf people access health care services and that the failure to provide the interpreters violates the rights of Deaf people. For an analysis of this decision from a critical disability studies approach, see Rioux and Valentine (2006); Pothier (2006).

10 On the tension between redistribution and recognition see Fraser (1998, 2000). On its implications for disability studies and activism, see Danermark and Gellerstedt (2004); Shakespeare (2006).

11 Obviously biological citizenship involves some form of mental or physical disability. Here I use the generic term 'disability activism' in reference to the various disability rights movements, which largely champion the social model of disability, discussed by disability studies scholars since the 1970s. On the history of disability rights activism, see Fleischer and Zames (2001).

12 NAAR and CAN have now merged their operations with Autism Speaks.

13 A good example of the extreme forms such criticism may take appears on the website Autistic.org, which shows a trash-can full of dead autistic foetuses labelled CAN placed in front of an abortion clinic, with the caption: 'The real meaning of "autism prevention"'. There is also a joke organisation known as CAN'T – Cure All NeuroTypicals – that parodies the name and acronym of Cure Autism Now – CAN.

14 Consequently, some self-advocates made a petition to the United Nations in 2004, asking to be recognised as a 'minority social group' deserving protection against 'discrimination' and 'inhuman treatment' (Nelson 2004).

15 These new transnational practices reproduce the same contradictions already perceived by Hannah Arendt between an abstract discourse on human rights and the need for a concrete national state backing them to become effective. Dezalay and Garth observe that: 'the movement for human rights is often presented as an exemplary illustration of those new transnational practices that escape from state order. However, by a sort of paradox, it is the national state's recognition of this "soft law" that represents the fulfilment of the militants' efforts, leading to a growing professionalization and competition within the market of political activism' (quoted in Rabinow 2005: 48).

16 For a short overview of the issue of human rights and disability see Bickenbach (2001).

17 Bickenbach (2001: 568) makes a useful classification of four basic types of legal expression of human rights for individuals with disabilities: enforceable anti-discrimination legislation; constitutional guarantees of equality; specific entitlement programmes; and voluntary human rights manifestos. They do not pretend to be mutually exclusive and are simply different vehicles by which to put human rights into work.

18 In relation to Canadian disability policy, Rioux and Valentine (2006: 48) verify that despite the Canadian Government's commitment to equality rights in different documents, 'substantive citizenship rights – especially at the provincial and municipal level – have not been attained in programs and services. Therefore,

most people with disabilities continue to be inhibited from achieving full citizenship'.

19 *Halderman v Pennhurst*, filed in Pennsylvania on behalf of the residents of the Pennhurst State School and Hospital denounced conditions at state schools for people with mental retardation. One resident of the Pennhurst State School and Hospital and her mother, later joined by other residents and the Parents and Family Association of Pennhurst, charged the School and Hospital with violating the rights of the patients and sought to improve their living conditions. The US district court, informed by the Developmental Disabilities Assistance and Bill of Rights Act, passed by the Congress in 1975, supported the position that institutionalisation was an infringement of constitutional rights and that they could receive the services in communal settings. Judge Broderick decided to move all the residents in communitarian settings. Although many parents did not applaud this decision, for they wanted the conditions at Pennhurst improved, Pennhurst would be closed in 1984. When the case was overturned at the US Supreme Court the decision was that, owing to the Eleventh Amendment, federal courts could not order state officials to conform to state laws.

20 See the section about 'The case of Christopher Reeve' in Baird, Rosenbaum and Toombs (2009), with articles by Goggin and Newell; Molsberry; Williams; Peace; and Schatzlein.

21 These issues have been vividly discussed in the literature. See, among others, Anstey (2002); Blume (1999, 2010); Brusky (1995); Campbell (2005a, 2005b, 2009); Davis (1997); Emery, Middleton and Turner (2010); Hyde and Power (2006); Häyry (2004, 2009); Kermit (2009); Key (1996); Lane and Grodin (1997); Levy (2002); McLellan (2002); Mundy (2002); Power (2005); Savulescu (2002); Sparrow (2005); Spriggs (2002); Tucker (1997, 1998); Young (2002).

22 This is not an easy choice, in cases of children more than 5 years old the success rate for adaptation to the cochlear implant drastically decreases. Children who remain for years without hearing do not adapt so well to the hearing device as those who start to use their residual hearing earlier (McLellan 2002).

23 When I speak here of the child's right to an open future and of maximising the choices of one's children I refer to the possibility of being part of the hearing world, the world of the larger society, as well as of the Deaf world, if he or she wishes. I am categorically neither referring nor defending the choice's ideology of late capitalism, according to which the more choices we have, the freer we are. Post-industrial society profits from the idea of choice. Consumer-citizens can choose not only among objects but also identities, gender, bodies and lifestyles. As Renata Salecl (2010) has convincingly argued, the impossible dream of total self-mastery underlying the overemphasis on choice does not bring happiness or tranquillity to the individual but exacerbates feelings of anxiety and insecurity. Choosing to be part of the Deaf community cannot be likened with the choosing of a lifestyle or a vacation destiny in the life-supermarket of the mainstream world.

24 Of course I do not want to overlook the fact that the actual technology is still imperfect, and some people have warned about the dangers of children remaining 'between cultures', not fully belonging to the hearing world and losing the bonds with the Deaf world. But even critics acknowledge that with technological development this problem could be overcome. Another objection concerns the

issue of whether somebody belonging to two cultures is not necessarily better off than individuals belonging to a single culture. Supporters of this argument highlight that normally the advantages of one culture will not be recognised by the other culture, especially if cultures are highly polarised owing to identity politics (Sparrow 2005). I certainly agree with this, but the issue at stake here is not one of 'bicultural' versus 'monocultural'. What is at stake is the child's right to an open future and not foreclosing his or her future options beforehand.

25 In *Bertinetti v Joy Mining Machinery*, 231 F. Supp. 2d 828 (S.D. Il. 2002), a boring mill operator with Charcot-Marie Tooth disease, a condition that affects the peripheral nervous system and extremities, filed an action under the ADA, claiming that the company failed reasonably to accommodate his disability. He argued that he was obliged to work at a specific mill that caused injuries to his back and legs. He also stated that the company had previously accommodated him by placing him in other mills, which implies an acknowledgement of his disabled condition. The district court did not accept the argument that the labourer was 'regarded as' disabled by virtue of evidence of the employer's efforts reasonably to accommodate his condition drawing from the Ninth Circuit Court of Appeals decision in *Thornton v McClatchy*, 261 F. 3d 789, 798 (9th Cir. 2001). The court acknowledged that he suffered from an 'impairment', although it did not believe that his limitations constituted a 'disability' under the ADA, for Bertinetti was not 'prevented' from walking. However, the court sustained Cox's argument that *Sutton* did not exclude individuals from the ADA if they refuse to take mitigating measures: 'A "disability" exists only where an impairment "substantially limits" a major life activity, not where it "might", "could", or "would" be substantially limiting if mitigating measures were not taken (*Sutton v United Air Lines, Inc.*, 527 U.S. 471, 482, 119 S.Ct. 2139, 144 L.Ed.2d 450 [1999]). Accordingly, this Court will not speculate about what Bertinetti's condition would be in the absence of mitigating measures that have actually been taken. On the other hand, the defendant urges the Court to consider that Bertinetti's condition might have been improved if he had utilized additional corrective devices'. *Sutton*, however, does not give courts a licence to 'meander in "would, could, or should-have" land' (*Nawrot v CPC Intern.*, 277 F.3d 896, 904 [7th Cir. 2002]). Courts should only consider mitigating steps actually taken and the consequences that actually followed.

26 About people living with the effects of thalidomide, see the interesting German documentary film *Nobody's Perfect* (2008) directed by Nico von Glasow.

27 There is a vast literature about the trial. See, among others, Baker (2006, 2007, 2011); Dawson (2004, 2005); Jackman (2007); Orsini (2009); Orsini and Smith (2010); Pothier (2006).

28 http://accessibilitynews.ca/cwdo/activities/health_committee.php?activities-health=411.

29 Both Locke in the 17th century and German legal thought in the 19th century posed the idea of the property of the body well before 20th century debates. Yet, such a theory had almost no practical consequences, since its applications concerned slavery, suicide and voluntary mutilations. These were the domains where legislation defined the status and protection of persons. The question of the distinction between the body and the person was never raised (Baud 1993).

30 The three cases involved children born with severe impairments who brought about a Wrongful Life cause of action in the New South Wales Supreme Court in 2002 (Campbell, F. 2009).

31 From a much quoted interview with Foucault (quoted in Bérubé 2010: 103).

32 In this sense Martha Nussbaum (2010) believes that the right to participate in the political process, the right to vote and other basic civil rights should be extended to individuals with severe cognitive disabilities on the basis of equality. The only way for those people not to be excluded from the basic functions of citizenship is to empower their guardians to exercise this function on that person's behalf and his or her interests. The idea 'of one person, one vote' should authorise the guardians or caregivers to vote in the interest of the person with disabilities. Without wanting to fetishise the act of voting as the best political action in advanced liberal societies, I believe Nussbaum's proposal constitutes a radical critique of liberal ideals of independence and autonomy and an important contribution to the undermining of the abstract disembodied person of liberal legalism.

Bibliography

Abi-Rached, J. (2008) 'The implications of the new brain sciences', *EMBO Reports*, 912: 1158–62.

____. and Rose, N. (2010) 'The birth of the neuromolecular gaze', *History of The Human Sciences*, 23: 11–36.

Adorno, T. and Horkheimer, M. (1947) *Dialectic of Enlightenment*, trans. J. Cumming, New York: Herder and Herder.

Agamben, G. (1998) *Homo Sacer: Sovereign Power and Bare Life*, trans. D. Heller-Roazen, Stanford: Stanford University Press.

Aïach, P. (1998) 'Les voies de la médicalisation', in P. Aïach and D. Delanoë (eds) *L'ère de la médicalisation. Ecce homo sanitas*, Paris: Ed. Economica.

Alaimo, D. (2008) 'Trans-Corporeal Feminisms and the Ethical Space of Nature', in S. Alaimo and S. Hekman (eds) *Material Feminisms*, Bloomington: Indiana University Press.

____. and Hekman S. (2008) 'Introduction: Emerging Models of Materiality in Feminist Theory', in S. Alaimo and S. Hekman (eds) *Material Feminisms*, Bloomington: Indiana University Press.

Andrieu, B. (2002a) *La nouvelle philosophie du corps*, Ramonville Saint-Agne: Éditions Érès.

____. (2002b) *La chair du cerveau. Phénoménologie et biologie de la cognition*, Belgique: Les Éditions de Sils Maria, Collection de Nouvelles Possibilités d'Existence.

____. (2003), *Le semaphore, naissance du sujet biotechnologique*, Belgique: Les Éditions de Sils Maria, Collection de Nouvelles Possibilités d'Existence.

____. (2004) 'Corps', in S. Leclercq (ed) *Abécédaire de Michel Foucault*, Paris: Les Belles Lettres.

Annoni, J. M. and Barras, V. (1993) 'La découpe du corps humain et ses justifications dans l'antiquité', *Canadian Bulletin for Medical History*, 10: 185–227.

Anstey, K. (2002) 'Are attempts to have impaired children justifiable?', *Journal of Medical Ethics*, 28: 286–88.

Antonetta, S. (2005) *A Mind Apart. Travels in a Neurodiverse World*, New York: Jeremy P. Tarcher/Penguin.

Appleby, J. (2002) 'Planned Obsolescence: Flying into the Future with Stelarc', in J. Zylinska (ed) *The Cyborg Experiments: The Extensions of the Body in the Media Age* (Technologies: Studies in Culture & Theory), London and New York: Continuum.

Arendt, H. (1978) *The Life of the Mind. Thinking*, New York: Harcourt Brace & Company.

____. (1979) *The Origins of Totalitarianism*, New York: Harcourt Brace & Company.
Aristotle (1995) *Parts of Animals*, in J. Barnes (ed) *The Complete Works of Aristotle*, Princeton: Princeton University Press.
Armstrong, R. (2002) 'Anger, Art and Medicine: Working with Orlan', in J. Zylinska (ed) *The Cyborg Experiments: The Extensions of the Body in the Media Age* (Technologies: Studies in Culture & Theory), London and New York: Continuum.
Assy, B. (2007) *Hannah Arendt: An Ethics of Personal Responsibility*. Hannah Arendt Studien, vol. 3. Edited by Antonia Grunenberg, Preface by Agnes Heller. 1st edn, Berlin, Oxford, New York and Frankfurt: Peter Lang.
Ausch, R., Doane, R. and Perez, L. (2000) 'Interview with Elizabeth Grosz', *Found Object*, 9: 1–16.
Ayers, R. (2000) 'Serene and Happy and Distant: An Interview with Orlan', in M. Featherstone (ed) *Body Modification*, London: Sage Publications.
Babo, M. A. (2004) 'Do corpo protésico ao corpo híbrido', *Revista de Comunicação e Linguagens*, 33: 25–35.
Badiou, A. (2004) *Le siècle*, Paris: Seuil.
Bagatell, N. (2007) 'Orchestrating voices: autism, identity and the power of discourse', *Disability & Society*, 22: 413–26.
Baird, R. M., Rosenbaum, S. E. and Toombs, S. K. (2009) 'The Case of Christopher Reeve', in R. M. Baird, S. E. Rosenbaum and S. K. Toombs (eds) *Disability: The Social, Political, and Ethical Debate*, Amherst and New York: Prometheus Books.
Baker, D. L. (2006) 'Autism as Public Policy', in D. Pothier and R. Devlin (eds) *Critical Disability Theory. Essays in Philosophy, Politics, Policy, and Law*, Vancouver: The University of British Columbia Press.
____. (2007) 'Defining autism in Canada: unfolding the public aspects of neurological disability', *The Social Science Journal*, 44: 687–97.
____. (2011) *The Politics of Neurodiversity. Why Public Policy Matters?* Boulder: Lynne Rienner.
Bakhtin, M. (1993) *Rabelais and His World*, trans. H. Iswolsky, Bloomington: Indiana University Press.
Balsamo, A. (1993) 'The virtual body in cyberspace', *Research in Philosophy and Technology*, 13: 119–39.
____. (1996) *Technologies of the Gendered Body. Reading Cyborg Women*, Durham: Duke University Press.
____. (1999) 'Forms of Technological Embodiment,' in J. Price and M. Shildrick (eds) *Feminist Theory and the Body: A Reader*, London: Routledge.
Barad, K. (2008) 'Posthumanist Performativity: Toward an Understanding of How Matter Comes to Matter', in S. Alaimo and S. Hekman (eds) *Material Feminisms*, Bloomington: Indiana University Press.
Barry, A., Osborne, T. and Rose, N. (eds) (1996) *Foucault and Political Reason. Liberalism, Neo-liberalism and Rationalities of Government*, Chicago: University of Chicago Press.
Baud, J. (1993) *L'affaire de la main volée. Une histoire juridique du corps*, Paris: Le Seuil.
Bauduin, T. M. (2012) 'Science, occultism, and the art of the Avant-Garde in the early twentieth century', *Journal of Religion in Europe*, 5: 23–55.
Beck, U. (1992) *Risk Society: Towards a New Modernity*, London: Sage Publications.
____., Giddens, A. and Lash, S. (1994) *Reflexive Modernization: Politics, Tradition and Aesthetics in the Modern Social Order*, Cambridge: Polity Press.

Bell, D. A. (1980) 'Brown v Board of Education and the Interest Convergence Dilemma', *Harvard Law Review*, 93: 518–33.

Belling, C. (1998) 'Reading the operation: television, realism, and the possession of medical knowledge', *Literature and Medicine*, 17: 1–23.

Bending, L. (2004) *The Representation of Bodily Pain in Late Nineteenth Century English Culture*, Oxford: Oxford University Press.

Benhabib, S. (1992) *Situating the Self: Gender, Community and Postmodernism in Contemporary Ethics*, New York: Routledge.

Benthien, C. and Wulf, C. (2001) 'Einleitung: Zur kulturellen Anatomie der Körperteile', in C. Benthien and C. Wulf (eds) *Körperteile. Eine Kulturelle Anatomie*, Hamburg: Rowohlt.

Bergmann, A. (1996) 'Die Verlebendigung des Todes und die Tötung des Lebendigen durch den medizinischen Bild', in E. Mixa et al (eds) *Körper-Geschlecht-Geschichte. Historische und Aktuelle Debatten in der Medizin*, Innsbruck-Wien: StudienVerlag.

Berlant, L. (1997) *The Queen of America Goes to Washington City*, Durham and London: Duke University Press.

Bermudez, J. L. (1998) 'Ecological Perception and the Notion of a Nonconceptual Point of View', in J. Bermudez, A. Marcel and N. Eilan (eds) *The Body and the Self*, Cambridge, Mass: MIT Press.

Berr, M. A. (1990) *Technik und Körper*, Berlin: Dietrich Reimer Verlag.

Bérubé, M. (2009) 'Citizenship and Disability', in R. M. Baird, S. E. Rosenbaum and S. K. Toombs (eds) *Disability: The Social, Political, and Ethical Debate*, Amherst and New York: Prometheus Books.

____. (2010) 'Equality, Freedom, and/or Justice for All: A Response to Martha Nussbaum', in E. F. Kittay and L. Carlson (eds) *Cognitive Disability and Its Challenge to Moral Philosophy*, Malden and Oxford: Wiley-Blackwell.

Bharadwaj, A. (2008) 'Biosociality and Biocrossing: Encounters with Assisted Conception and Embryonic Stem Cells in India', in S. Gibbon and C. Novas (eds) *Biosocialities, Genetics and the Social Sciences. Making Biological Identities*, London and New York: Routledge.

Bickenbach, J. E. (2001) 'Disability Human Rights, Law and Policy', in G. Albrecht, K. Seelman and M. Bury (eds) *Handbook of Disability Studies*, Thousand Oaks, London and New Delhi: Sage Publications.

Biehl, J. (2004) 'The activist state: global pharmaceuticals, AIDS, and citizenship in Brazil', *Social Text*, 22: 105–32.

____. (2006) 'Will to live: AIDS drugs and local economies of salvation', *Public Culture*, 18: 457–72.

Binkley, S. and Capetillo-Ponce, J. (eds) (2010) *A Foucault for the 21st Century: Governmentality, Biopolitics, and Discipline in the New Millennium*, Newcastle: Cambridge Scholars Publishing.

Blackman, L. (2008) *The Body: The Key Concepts*, Oxford: Berg.

Bleeker, M. (ed) (2008) *Anatomy Live: Performance and the Operating Theatre*, Amsterdam: Amsterdam University Press.

Blume, H. (1997) 'Autism and the internet, or, It's the wiring, stupid', available at http://web.mit.edu/comm-forum/papers/blume.html (accessed 20 June 2007).

Blume, S. S. (1992) *Insight and Industry, On the Dynamics of Technological Change in Medicine*, Cambridge, Mass and London: MIT Press.

____. (1999) 'Histories of cochlear implantation', *Social Science & Medicine*, 49: 1257–68.

____. (2006) 'Anti-vaccination movements and their interpretations', *Social Science & Medicine*, 62: 628–42.

____. (2010) *The Artificial Ear: Cochlear Implants and the Culture of Deafness*, New Brunswick, NJ and London: Rutgers University Press.

Blumenberg, H. (1993) 'Light as Metaphor for Truth: At the Preliminary Stage of Philosophical Concept Formation', in D. M. Levin (ed) *Modernity and the Hegemony of Vision*, Berkeley and Los Angeles: University of California Press.

Boccioni, U. et al (1910) *Manifesto tecnico della pittura futurista*, 11 April 1910, available at http://www.futurismo1.com/paint/maniftecn.htm (accessed 12 November 2008).

Borck, C. (2001) 'Die Unhintergehbarkeit des Bildschirms. Beobachtungen zur Rolle von Bildtechniken in den präsentierten Wissenschaften', in B. Heinz and J. Huber (eds) *Mit dem Auge denken. Strategien der Sichtbarmachung in Wissenschaftlichen und Virtuellen Welten*, Zürich: Edition Voldemeer.

Bordo, S. (1993) *Unbearable Weight: Feminism, Western Culture, and the Body*, Berkeley and Los Angeles: University of California Press.

____. (1998) 'Bringing Body to Theory', in D. Welton (ed) *Body and Flesh: A Philosophical Reader*, Malden: Blackwell Publishers Ltd.

Borrillo, D. (1994) 'Estatuto y representacion del cuerpo humano en el sistema juridico', *Revista Española de Investigaciones Sociológicas*, 68: 211–22.

Bostrom, N. (2005) 'A history of transhumanist thought', *Journal of Evolution and Technology*, 14: 1–25.

____. (2009) 'Why I Want to Be a Posthuman When I Grow Up', in B. Gordijn and R. Chadwick (eds) *Medical Enhancement and Posthumanity*, Oxford: Oxford University Press.

Bourdieu, P. (1980) *Le sens pratique*, Paris: Les Éditions de Minuit.

____. (2010) *Distinction. A Social Critique of Judgment of Taste*, trans. R. Nice, Abingdon, Oxon: Routledge.

Bourke, J. (2012) 'Languages of pain', *The Lancet*, 379: 2420–21.

Braidotti, R. (2002) *Metamorphoses: Towards a Feminist Theory of Becoming*, Cambridge: Polity Press.

Brandt, A. and Rozin, P. (eds) (1997) *Morality and Health*, London: Routledge.

Broadhurst, S. and Machon, J. (eds) (2009) *Sensualities/Textualities and Technologies: Writings of the Body in 21st Century Performance*, Basingstoke and New York: Palgrave Macmillan.

Bröckling, U., Krasmann, S. and Lemke, T. (eds) (2011) *Governmentality. Current Issues and Future Challenges*, New York: Routledge.

Brodie, J. M. (2002) 'Three Stories of Canadian Citizenship', in R. Adamoski, D. E. Chunn and R. Menzies (eds) *Constructing Canadian Citizenship: Historical Readings*, Peterborough: Broadview Press.

Brooks, R. A. (2003) *Flesh and Machines. How Robots Will Change Us*, New York: Vintage Books.

Brown, N. and Michael, M. (2003) 'A sociology of expectations: retrospecting prospects and prospecting retrospects', *Technology Analysis & Strategic Management*, 15: 3–18.

Brown, P. (1978) *The Making of Late Antiquity*, Cambridge, Mass: Harvard University Press.

____. (1982) 'The Rise and Function of the Holy Man in Late Antiquity', *Society and the Holy in Late Antiquity*, Berkeley: University of California Press.

____. (2008) *The Body and Society. Men, Women, & Sexual Renunciation in Early Christianity*, Twentieth-Anniversary edn, New York: Columbia University Press.

Brusky, A. E. (1995) 'Making decisions for deaf children regarding cochlear implants: the legal ramifications of recognizing deafness as a culture rather than a disability', *Wisconsin Law Review*, 1: 235–70.

Bülow, P. H. (2008) 'Tracing Contours of Contestation in Narratives about Chronic Fatigue Syndrome', in P. Moss and K. Teghtsoonian (eds) *Contesting Illness: Processes and Practices*, Toronto, Buffalo and London: University of Toronto Press.

Bunton, R. (1997) 'Popular Health, Advanced Liberalism and *Good Housekeeping*', in A. Petersen and R. Bunton (eds) *Foucault, Health and Medicine*, London: Routledge.

Burchell, G., Gordon, C. and Miller, P. (eds) (1991). *The Foucault Effect. Studies in Governmentality*, Hemel Hempstead: Harvester Wheatsheaf.

Burke, L. (2008) 'The country of my disease: genes and genealogy in Alzheimer's life-writing', *Journal of Literary & Cultural Disability Studies*, 2: 63–74.

Burri, R. V. and Dumit, J. (2008) 'Social Studies of Scientific Imaging and Visualization', in E. J. Hackett, O. Amsterdamska and W. Lynch (eds) *The Handbook of Science and Technology Studies*, 3rd edn, Cambridge, Mass: MIT Press.

Burwood, S. (2008) 'The apparent truth of dualism and the uncanny body', *Phenomenology and Cognitive Sciences*, 7: 263–78.

Butler, J. (1989) 'Foucault and the paradox of bodily inscriptions', *The Journal of Philosophy*, 86: 601–607.

____. (1990) *Gender Trouble: Feminism and the Subversion of Identity*, New York: Routledge.

____. (1997) 'How can I deny that these hands and body are mine?' *Qui parle*, 11: 1–20.

____. (2011) *Bodies That Matter. On the Discursive Limits of 'Sex'*, London and New York: Routledge.

Butterworth, G. (1998) 'An Ecological Perspective on the Origins of Self', in J. Bermudez, A. Marcel and N. Eilan (eds) *The Body and the Self*, Cambridge, Mass: MIT Press.

Bynum, C. (1995) 'Why all the fuss about the body? A medievalist's perspective', *Critical Inquiry*, 22: 1–33.

Callon, M. and Rabeharisoa, V. (2004) 'Gino's lesson on humanity: genetics, mutual entanglements and the sociologist's role', *Economy and Society*, 33: 1–27.

Cameron, A. (2002) 'Ascetic Closure and the End of Antiquity', in V. L. Wimbush and R. Valantasis (eds) *Asceticism*, New York: Oxford University Press.

____. (2012) *The Mediterranean World in Late Antiquity: AD 395–700* (The Routledge History of the Ancient World) 2nd edn, London and New York: Routledge.

Campbell, A. V. (2009) *The Body in Bioethics*, London and New York: Routledge.

Campbell, F. K. (2005a) 'Legislating Disability: Negative Ontologies and the Government of Legal Identities', in S. Tremain (ed) *Foucault and the Government of Disability*, Ann Arbor: The University of Michigan Press.

____. (2005b) 'Selling the cochlear implant', *Disability Studies Quarterly* (online version), 25(3).

____. (2009) *Contours of Ableism: The Production of Disability and Abledness*, Basingstoke: Palgrave Macmillan.

Campbell, I. (1998) 'For Esmé with Love and Squalor', in J. Campbell and J. Harbord (eds) *Psychopolitics and Cultural Desires*, London and New York: University College Press.

Campos, P. et al (2006) 'The epidemiology of overweight and obesity: public health crisis or moral panic?', *International Journal of Epidemiology*, 35: 55–60.

Canguilhem, G. (1989) 'Le normal et le pathologique', in *La connaissance de vie*, Paris: Vrin.

____. (1991) *The Normal and the Pathological*, trans R. Fawcett and Robert S. Cohen, New York: Zone Books.

____. (1997) 'On *Histoire de la Folie* as Event', in A. Davidson (ed) *Foucault and His Interlocutors*, Chicago: University of Chicago Press.

Capelle, W. (1967) 'Asceticism (Greek)', in J. Hastings (ed) *Encyclopaedia of Religion and Ethics*, vol. II, Edinburgh: T. & T. Clark.

Caplan, A. (2005) 'Would you have allowed Bill Gates to be born? Advances in prenatal genetic testing pose tough questions', available at http://www.nbcnews.com/id/7899821/ns/health-health_care/t/would-you-have-allowed-bill-gates-be-born/ (accessed 31 May 2005).

Carey, A. C. (2009) *On the Margins of Citizenship: Intellectual Disability and Civil Rights in Twentieth Century America*, Philadelphia: Temple University Press.

Carey, K. N. (2005) 'Wrongful life and wrongful birth: legal aspects of failed genetic testing in oocyte donation', *Penn Bioethics Journal*, 1: 1–4.

Carlino, A. (1999) *Books of the Body. Anatomical Ritual and Renaissance Learning*, trans. John Tedeschi and Anne C. Tedeschi, Chicago and London: University of Chicago Press.

Carson, G (1957) *Cornflakes Crusade*, New York: Reinhart.

Cartwright, L. (1997) *Screening the Body. Tracing Medicine's Visual Culture*, Minneapolis: University of Minnesota Press.

Castagno, A. E. and Lee, S. J. (2007) 'Native mascots and ethnic fraud in higher education: using tribal critical race theory and the interest convergence principle as an analytic tool', *Equity and Excellence in Education*, 40: 3–13.

CBS. (2006) 'Retraining the brain: doctors test drug-free methods to restore lost mental capabilities', *CBS NEWS*, 15 January, available at http://www.cbsnews.com/stories/2006/01/15/sunday/main1209916.shtml (accessed 15 January 2006).

Chafetz, M. D. (1992) *Smart for Life: How to Improve Your Brain Power at Any Age*, New York: Penguin.

Chamak, B. (2008) 'Autism and social movements: French parents' associations and international autistic individuals' organisations', *Sociology of Health and Illness*, 30: 76–96.

Cheah, P., Fraser, D. and Grbich, J. (1996) 'Introduction: The Body of the Law', in P. Cheah, D. Fraser and J. Grbich (eds) *Thinking Through the Body of the Law*, New York: New York University Press.

Cheah, P. and Grosz, E. (1996) 'The body of the law: notes toward a theory of corporeal justice', in P. Cheah, D. Fraser and J. Grbich (eds) *Thinking Through the Body of the Law*, New York: New York University Press.

Cheek, J. (2008) 'Healthism: a new conservatism?', *Qualitative Health Research*, 18: 974–82.

Cheu, J. (2004) 'De-gene-erates, Replicants and Other Aliens: (Re)defining Disability in Futuristic Film', in M. Corker and T. Shakespeare (eds) *Disability/Postmodernity: Embodying Disability Theory*, London: Continuum.

Clark, A. (2003) *Natural-Born Cyborgs. Minds, Technologies and the Future of Human Intelligence*, New York: Oxford University Press.

Clark, E. A. (1999) *Reading Renunciation: Asceticism and Scripture in Early Christianity*, Princeton: Princeton University Press.

Clarke, A. E. et al (2010a) 'Biomedicalization. A Theoretical and Substantive Introduction', in A. Clarke, L. Mamo, J. Ruth Foster, J. R. Fishman and J. K. Shim (eds) *Biomedicalization. Technoscience, Health and Illness in the U.S.*, Durham and London: Duke University Press.

____. (2010b) 'Biomedicalization. Technoscientific Transformations of Health, Illness, and U.S. Biomedicine', in A. Clarke, L. Mamo, J. Ruth Foster, J. R. Fishman and J. K. Shim (eds) *Biomedicalization. Technoscience, Health and Illness in the U.S.*, Durham and London: Duke University Press.

Clarke, J. (2000) 'The Sacrificial Body of Orlan', in M. Featherstone (ed) *Body Modification*, London: Sage Publications.

____. (2002) 'The Human/Not Human in the Work of Orlan and Stelarc', in J. Zylinska (ed) *The Cyborg Experiments: The Extensions of the Body in the Media Age*, London and New York: Continuum.

Clarke, J. and van Amerom, G. (2007) '"Surplus Suffering": differences between organizational understandings of Asperger's syndrome and those people who claim the "disorder"', *Disability & Society*, 22: 761–76.

Clarke, J. H. et al (2007) *Creating Citizen Consumers: Changing Publics and Changing Public Services*, London: Sage Publications.

Clarke, L. H. and Bennett, E. V. (2012) 'Constructing the moral body: self-care among older adults with multiple chronic conditions', *Health*, 17: 211–28.

Cleland, K. (2010) 'Prosthetic bodies and virtual cyborgs,' *Second Nature*, 3: 74–101.

Clough, P. T. (2012) 'Feminist Theory: Bodies, Science and Technology', in B. S. Turner (ed) *Routledge Handbook of Body Studies*, London: Routledge.

Cohen, I. and Goldsmith, M. (2002) *Hands On. How to Use Brain Gym in the Classroom*, Ventura, CA: Edu Kinesthetics, Inc.

Cohen, R. (1984) 'Merleau-Ponty, the flesh and Foucault', *Philosophy Today*, 28: 329–38.

Cohn, S. (2008) 'Making objective facts from intimate relations: the case of neuroscience and its entanglements with volunteers', *History of the Human Sciences*, 21: 86–103.

Colebrook, C. (2008) 'On not Becoming Man: The Materialist Politics of Unactualized Potential', in S. Alaimo and S. Hekman (eds) *Material Feminisms*, Bloomington: Indiana University Press.

Connor, J. (2007) 'Exhibit essay review: "Faux Reality" Show? The Body Worlds phenomenon and its reinvention of anatomical spectacle,' *Bulletin of the History of Medicine*, 81: 848–62.

Conrad, J. (1995) 'Egyptian and Syrian asceticism in late antiquity', *Ex Post Facto*, 4: np, available at http://sfsu.edu/~hsa/ex-post-facto/ascetic.html (accessed 12 July 2009).

Conrad, P. (1995) 'Médicalisation et contrôle social', in L. Bouchard and D. Cohen (eds) *Médicalisation et contrôle social*, Quebec: Les cahiers scientifiques de l'Acfas, 84.

——. (2007) *The Medicalization of Society. On the Transformation of Human Conditions into Treatable Disorders*, Baltimore: Johns Hopkins University Press.
Consumer Citizenship Network (CCN). (2005) *Consumer Citizenship Education – Guidelines, Higher Education*, V. W. Thoresen (ed), Hamar: CNN.
Cook, H. (2006) 'Medicine', in K. Park and L. Daston (eds) *The Cambridge History of Science, Volume 3: Early Modern Science*, Cambridge: Cambridge University Press.
Cooper, M. (2008). *Life As Surplus: Biotechnology and Capitalism in the Neoliberal Era*, Seattle and London: University of Washington Press.
Cooter, R. (1984) *The Cultural Meaning of Popular Science. Phrenology and the Organisation of Consent in 19th Century Britain*, Cambridge: Cambridge University Press.
Cooter, R. (2008) 'Biocitizenship', *The Lancet*, 372: 1725.
Cooter, R. (2010) 'The turn of the body: history and the politics of the corporeal', *ARBOR – Ciencia, Pensamiento y Cultura*, 186: 393–405.
____. and Pickstone, J. (2003) (eds) *Companion Encyclopedia of Medicine in the Twentieth Century*, London and New York: Routledge.
____. and Stein, C. (2010) 'Cracking biopower', *History of the Human Sciences*, 23: 109–28.
Costa, J. F. (2004) *O vestígio e a aura. Corpo e consumismo na moral do espetáculo*, Rio de Janeiro: Garamond.
Cottingham, J. (1986) *Descartes*, Oxford: Basil Blackwell.
Courtine, J. J. (1993) 'Les stakhanovistes du narcissisme. Body-building et puritanisme ostentatoire dans la culture américaine du corps', *Communications*, 56: 225–51.
Coveney, J. (2006) *Food, Morals and Meaning. The Pleasure and Anxiety of Eating*, 2nd edn, London and New York: Routledge.
Covino, D. C. (2004) *Amending the Abject Body: Aesthetic Makeovers in Medicine and Culture*, SUNY Seried in Feminist Criticism and Theory, Albany: State University of New York Press.
Cox, J. (2009) '"Corrective" surgery and the Americans with Disabilities Act', *San Diego Law Review*, 46, available at http://papers.ssrn.com/sol3/papers.cfm?abstract_id=1274285 (accessed 29 April 2009).
Craig, J. R. (2002) 'Buddhism, science, and market: the globalisation of Tibetan medicine', *Anthropology & Medicine*, 9: 267–89.
Crawford, R. (1980) 'Healthism and the medicalization of everyday life', *International Journal of Health Services*, 10: 365–88.
____. (1984) 'A Cultural Account of Health', in J. McKinlay (ed) *Issues in the Political Economy of Health Care*, London: Tavistock.
____. (1994) 'The boundaries of the self and the unhealthy other: reflections on health, culture and aids', *Social Sciences and Medicine*, 38(10): 1347–65.
____. (2004) 'Risk ritual and the management of control and anxiety in medical culture,' *Health*, 8: 505–28.
____. (2006) 'Health as a meaningful social practice', *Health*, 10: 401–20.
Crease, R. (1993) 'Biomedicine in the Age of Imaging', *Science*', 261: 554–61.
Crossley, N. (2001) *The Social Body. Habit, Identity and Desire*, London: Sage Publications.
____. (2004) 'The circuit trainer's habitus: reflexive body techniques and the sociality of the workout', *Body & Society*, 10: 37–69.
Cruz, R. (2012) 'Mutations and metamorphoses: body horror is biological horror', *Journal of Popular Film and Television*, 40: 160–68.

Csordas, T. J. (2000) 'Computerized Cadavers: Shades of Being and Representation in Virtual Reality', in P. E. Brodwin (ed) *Biotechnology and Culture, Bodies, Anxieties, Ethics*, Bloomington and Indianapolis: Indiana University Press.

____. (2001) 'Introduction: The Body as Representation and Being-in-the-world', in T. J. Csordas (ed) *Embodiment and Experience. The Existential Ground of Culture and Self*, Cambridge: Cambridge University Press.

Curry, L. (2002) *Human Body on Trial. A Sourcebook with Cases, Laws, and Documents*, Indianapolis: Hackett Publishing Company.

Dalton, D. (2012) 'The object of anxiety: Heidegger, Levinas and the phenomenology of the dead', *Janus Head: Interdisciplinary Studies in Literature, Continental Philosophy, Phenomenological Psychology, and the Arts*, Special Issue, 12: 67–82.

Danermark, B. and Coniavitis Gellerstedt, L. (2004) 'Social justice: redistribution and recognition – a non-reductionist perspective on disability', *Disability and Society*, 19: 339–53.

Danius, S. (2000), 'Novel visions and the crisis of culture: visual technology, modernism, and death in *The Magic Mountain*', *Boundary*, 2: 177–211.

Daston, L. and Galison, P. (1992) 'The image of objectivity', *Representations*, 40: 81–128.

____. (2007) *Objectivity*, New York: Zone Books.

Davidson, J. and Henderson, V. L. (2010) '"Coming out" on the spectrum: autism, identity and disclosure', *Social & Cultural Geography*, 11: 155–70.

Davis, D. S. (1997) 'Genetic dilemmas and the child's right to an open future', *Hastings Center Report*, 27: 7–15.

Dawson, M (2004) 'Factum of the intervener', available at http://www.sentex.net/~nexus23/naa_fac.html (accessed 10 July 2008).

____. (2005) 'An autistic victory: the true meaning of the Auton decision', available at http://www.sentex.net/~nexus23/naa_vic.html (accessed 10 July 2008).

de Rijcke, S. and Beaulieu, A. (2007) 'Taking a good look at why scientific images don't speak for themselves', *Theory & Psychology*, 17: 733–42.

Dean, M. (1994) *Critical and Effective Histories: Foucault's Methods and Historical Sociology*, London and New York: Routledge.

____. (1996) 'Foucault, Government and the Enfolding of Authority', in A. Barry, T. Osborne and N. Rose (eds) *Foucault and Political Reason. Liberalism, Neo-liberalism and Rationalities of Government*, Chicago and London: University of Chicago Press.

Debord, G. (1999) *The Society of the Spectacle*, trans. D. Nicholson-Smith, New York: Zone Books.

DeHart, S. M. (2000) 'Hippocratic medicine and the Greek body image', *Perspectives on Science*, 7: 349–82.

Deleuze, G. and Guattari, F. (1987) *A Thousand Plateaus*, trans. B. Massumi, Minneapolis: University of Minnesota Press.

Denier, Y. (2007) *Efficiency, Justice and Care. Philosophical Reflections on Scarcity in Health Care*, Dordrecht: Springer.

Dennison, E. P. and Dennison, E. G. (1989) *Brain Gym*, teachers' edn, Ventura, CA: Edu-Kinesthetics.

Depraz, N. (1997) 'La traduction de Leib: une crux phaenomenologica', *Etudes phénoménologiques*, 26: 91–109.

Derrida, J. (1992) 'Force of Law: The "Mystical Foundation of Authority"', in D. Cornell and M. Rosenfeld (eds) *Deconstruction and the Possibility of Justice*, New York: Routledge.

Descartes, R. (1996) [1647] *Méditations on First Philosophy*, in C Adam and P. Tannery (eds) *Oeuvres de Descartes*, Paris: Vrin.

____. (1996) [1639] *Correspondance, Mars 1638–Décembre 1639*, in C Adam and P. Tannery (eds) *Oeuvres de Descartes*, Paris: Vrin.

Detel, W. (2005) *Foucault and Classical Antiquity: Power, Ethics and Knowledge*, trans. D.Wigg-Wold, Cambridge: Cambridge University Press.

Devlin, R. and Pothier, D. (2006) 'Introduction: Toward a Critical Theory of Discitizenship', in D. Pothier and R. Devlin (eds) *Critical Disability Theory. Essays in Philosophy, Politics, Policy, and Law*, Vancouver: University of British Columbia Press.

Dewey, J (1989) [1929] *Experience and Nature*, La Salle, Il: Open Court.

Diprose, R. (2012) 'Continental philosophy: thinking the corporeal with the political', *The Southern Journal of Philosophy*, 50: 220–33.

Dolby, T. and Alker, G. (1997) *Origins and Development of Medical Imaging*, Carbondale and Edwardsville: Southern Illinois University Press.

Douzinas, C. (2008) *Human Rights and Empire. The Political Philosophy of Cosmopolitanism*, Abingdon and New York: Routledge.

——. and Gearey, A. (2005) *Critical Jurisprudence. The Political Philosophy of Justice*, Oxford and Portland: Hart.

Doyle, J. (2007) 'Cybersurgery and surgical (dis)embodiment: Technology, science, art and the body'. Special Issue: 'Walter Benjamin and the Virtual: Politics, Art, and Mediation in the Age of Global Culture', available at http://www.transformationsjournal.org/journal/issue_15/article_03.shtml (accessed 15 November 2010).

Dreyfus, H. L. and Rabinow, P. (1983) *Michel Foucault. Beyond Structuralism and Hermeneutics*, Chicago: University of Chicago Press.

Drijvers, H. J. W. (1984) 'Athleten des Geistes. Zur politischen Rolle der syrischen Asketen und Gnostiker', in J. Taubes (ed) *Religionstheorie und Politische Theologie. Gnosis und Politik*, München: Wilhelm Fink Verlag.

____. (1985) 'Askese und Mönchtum im frühen Christentum', in W. Schluchter (ed) *Max Webers Sicht des Antiken Christentums*, Frankfurt: Suhrkamp.

Druhle, M. and Clement, S. (1998) 'Enjeux et formes de la médicalisation: d'une approche globale au cas de la gérontologie', in P. Aïach and D. Delanoë (eds) *L'ère de la médicalisation. Ecce homo sanitas*, Paris: Ed. Economica.

Duden, B. (1993) 'Die Frau ohne Unterleib: Zu Judith Butler Entkörperung', *Feministische Studien*, 2: 24–33.

——. (1997) *The Woman Beneath the Skin. A Doctor's Patients in Eighteenth-Century Germany*, trans. Thomas Dunlap, Cambridge, Mass: Harvard University Press.

Dumit, J. (2004) *Picturing Personhood. Brain Scans and Biomedical Identity*, Princeton: Princeton University Press.

____. (2006) 'Illnesses you have to fight to get: facts as forces in uncertain, emergent illnesses', *Social Science & Medicine*, 62: 577–90.

Dumont, L. (1986) *Essays on Individualism: Modern Ideology in Anthropological Perspective*, Chicago: University of Chicago Press.

Duranti, A. (2010) 'Husserl, intersubjectivity and anthropology', *Anthropological Theory*, 10: 16–35.

Durbach, N. (2012) '"Skinless Wonders": Body Worlds and the Victorian Freak Show', *Journal of the History of Medicine and Allied Sciences*, June, available at http://

jhmas.oxfordjournals.org/content/early/2012/06/01/jhmas.jrs035.full.pdf+html (accessed 10 June 2012).

Edgley, C. (2006) 'The Fit and Healthy Body: Consumer Narratives and the Management of Postmodern Corporeity', in D. Waskul and P. Vannini (ed) *Body/Embodiment: Symbolic Interaction and the Sociology of the Body*. Burlington, VT: Ashgate.

____. and Brisset, D. (1990) 'Health Nazis and the cult of the perfect body: some polemical observations', *Symbolic Interaction*, 2(13): 257–79.

Ehrenberg, A. (2004) 'Le sujét cerebral', *Esprit*, 309: 130–55.

Eiffert, S. D. (1999) *Cross-train Your Brain: A Mental Fitness Program for Maximizing Creativity and Achieving Success*, New York: Amacom.

Elias, N. (1978) [1939] 'The History of Manners', in E. Dunning, J. Goudsblom and S. Mennell (eds) *The Civilizing Process*, Oxford: Blackwell.

Elliott, C. (2003) *Better than Well. American Medicine Meets the American Dream*, New York and London: W. W. Norton & Company.

Emery, S. D., Middleton, A. and Turner, G. H. (2010) 'Whose deaf genes are they anyway? The deaf community's challenge to legislation on embryo selection', *Sign Language Studies*, 10: 155–69.

Engel, D. M. and Munger, F. W. (2003) *Rights of Inclusion: Law and Identity in the Life Stories of Americans With Disabilities*, Chicago and London: University of Chicago Press.

Epstein, S. (2008) 'Patient Groups and Health Movements', in M. Lynch and J. Wajcman (eds) *The Handbook of Science and Technology Studies*, 3rd edn, Cambridge, MA: MIT Press.

Farnell, R. (2000) 'In Dialogue with 'Posthuman' Bodies: Interview with Stelarc', in M. Featherstone (ed) *Body Modification*, London: Sage Publications.

Fassin, D. (2011) *Humanitarian Reason. A Moral History of the Present*, trans. R. Gomme, Berkeley, Los Angeles and London: University of California Press.

Faure, O. (1998) 'La médicalisation vue par les historiens', in P. Aïach and D . Delanoë (eds) *L'ère de la médicalisation. Ecce homo sanitas*, Paris: Ed. Economica.

Favazza, A. R. (1996) *Bodies Under Siege. Self-mutilation and Body Modification in Culture and Psychology*, Baltimore and London: Johns Hopkins University Press.

Featherstone, M. (2000) 'Body Modification: An Introduction', in M. Featherstone (ed) *Body Modification*, London: Sage Publications.

Feher, M. (ed) (1989), *Fragments for a History of the Human Body*, vol. 3, New York: Zone Books.

Ferrari, G. (1987) 'Public anatomy lessons and the carnival: the anatomical theatre of Bologna', *Past and Present*, 117(1): 50–106.

Fisher, J. A. (2002) 'Tattooing the body, marking culture', *Body and Society*, 8: 91–107.

Fitzgerald, R. (2008) 'Biological citizenship at the periphery: parenting children with genetic disorders', *New Genetics & Society*, 27: 251–66.

Flear, M. L. (2008) '"Together for health?" How EU governance of health undermines active biological citizenship', *Wisconsin International Law Journal*, 26: 868–907.

Fleischer, D. Z. and Zames, F. (2001) *The Disability Rights Movement. From Charity to Confrontation*, Philadelphia: Temple University Press.

Foucault, M. (1978) *The History of Sexuality: An Introduction*, trans. R. Hurley, Harmondsworth: Penguin.

——. (1980) *Power Knowledge. Selected Interviews & other Writings 1972–1977*. Edited by C. Gordon. New York: Pantheon Books.
——. (1983) 'Afterword: The Subject and Power', in H. F. Dreyfus and P. Rabinow (eds) *Michel Foucault: Beyond Structuralism and Hermeneutics*, Brighton: Harvester Press.
——. (1986) *The Care of the Self. Volume 3 of The History of Sexuality*, trans. R. Hurley, New York: Pantheon Books.
——. (1988) 'Technologies of the self', in L. H. Martin, H. Gutman, and P. H. Hutton (eds) *Technologies of the self*, Amherst: University of Massachusetts Press.
——. (1990) *The Use of Pleasure. Volume 2 of The History of Sexuality*, trans. R. Hurley, New York: Vintage Books.
____. (1991) *Remarks on Marx. Conversations With Duccio Trombadori*. trans. James Goldstein and James Cascaito, New York: Semiotext(e).
——. (1994a) *The Order of Things. An Archeology of the Human Sciences*, trans. A. Sheridan, New York: Vintage Books.
——. (1994b) *The Birth of the Clinic. An Archaelogy of Medical Perception*, trans. A. Sheridan, New York: Vintage Books.
——. (1995), *Discipline and Punish: the Birth of the Prison*, trans. A. Sheridan, New York: Vintage Books.
____. (1997a) 'The Ethics of the Concern for Self as a Practice of Freedom', in P. Rabinow (ed) *The Essential Works of Michel Foucault 1954–1984: Volume One. Ethics, Subjectivity and Truth*, New York: The New Press.
____. (1997b) 'Sexual Choice, Sexual Act', in P. Rabinow (ed) *The Essential Works of Michel Foucault 1954–1984: Volume One. Ethics, Subjectivity and Truth*, New York: The New Press.
____. (1997c) 'The Social Triumph of the Sexual Will', in P. Rabinow (ed) *The Essential Works of Michel Foucault 1954–1984: Volume One. Ethics, Subjectivity and Truth*, New York: The New Press.
____. (1997d) 'Michel Foucault, An Interview: Sex, Power and the Politics of Identity', in P. Rabinow (ed) *The Essential Works of Michel Foucault 1954–1984: Volume One. Ethics, Subjectivity and Truth*, New York: The New Press.
____. (1997e) 'On the Genealogy of Ethics: An Overview of Work in Progress', in P. Rabinow (ed) *The Essential Works of Michel Foucault 1954–1984: Volume One. Ethics, Subjectivity and Truth*, New York: The New Press.
——. (1998) 'Nietzsche, Genealogy, History', in James D. Faubon (ed) *The Essential Works of Foucault 1954–1984, Volume Two. Aesthetics, Method, and Epistemology*, New York: The New Press.
____. (1999) 'Michel Foucault and Zen: A Stay in a Zen Temple', in Jeremy R. Carrette (ed) *Religion* and *Culture*: *Michel Foucault*, New York: Routledge.
——. (2001) *Fearless Speech*, Los Angeles: Semiotext(e).
——. (2005) *The Hermeneutics of the Subject. Lectures at the Collège de France, 1981–1982*, trans. G. Burchell, New York: Palgrave.
____. (2006a) *Psychiatric Power*: *Lectures at the Collège de France, 1973–1974*, trans. G. Burchell, Basingstoke and New York: Palgrave Macmillan.
____. (2006b) 'Utopian Body', in C. A. Jones (ed) *Sensorium: Embodied Experience, Technology, and Contemporary Art*, Cambridge, MA: MIT Press.
Foxhall, L. (1998) 'Pandora Unbound: A Feminist Critique of Foucault's History of Sexuality', in H. David, J. Larmour, P. A. Miller and C. Platter (eds) *Rethinking*

Sexuality: Foucault and Classical Antiquity, Princeton: Princeton University Press.
Fraser, N. (1998) 'Social Justice in the Age of Identity Politics: Redistribution, Recognition and Participation', in G. Peterson (ed) *The Tanner Lectures on Human Values*, vol. 19, Salt Lake City: Utah University Press.
____. (2000) 'Rethinking recognition', *New Left Review*, 3: 107–20.
Freeman, S. (2003) *Science for Sale in the Autism Wars*, Langley: SFK Books.
French, R. (1994) 'The Anatomical Tradition', in W. F. Bynum and R. Porter (eds) *Companion Encyclopedia of the History of Medicine*, London and New York: Routledge.
Freud, S. (1919) 'The "Uncanny"', in J. Strachey (ed) *The Standard Edition of the Complete Psychological Works of Sigmund Freud, Volume XVII (1917–1919): An Infantile Neurosis and Other Works*, London: Hogarth.
Friedner, M. (2010) 'Biopower, biosociality, and community formation: how biopower is constitutive of the deaf community', *Sign Language Studies*, 10: 336–47.
Gallagher, S and Zahavi, D. (2008) *The Phenomenological Mind: An Introduction to Philosophy of Mind and Cognitive Science*, Abingdon and New York: Routledge.
Galvin, R. (2002) 'Disturbing notions of chronic illness and individual responsibility: towards a genealogy of morals', *Health*, 6: 107–37.
Geary, J. (2002) *The Body Electric. An Anatomy of the New Bionic Senses*, London: Weidenfeld & Nicolson.
Gibbon, S. (2008) 'Charity, Breast Cancer Activism and the Iconic Figure of the BRCA Carrier', in S. Gibbon and C. Novas (eds) *Biosocialities, Genetics and the Social Sciences. Making Biologies and Identities*, London and New York: Routledge.
____. and Novas, C. (2008) 'Biosocialities, Genetics and the Social Sciences', in S. Gibbon and C. Novas (eds) *Biosocialities, Genetics and the Social Sciences. Making Biologies and Identities*, London and New York: Routledge.
Gibson, W. (1984) *Neuromancer*, New York: Ace Books.
Giddens, A. (1991) *Modernity and Self-identity: Self and Society in the Late Modern Age*, Stanford: Stanford University Press.
Gilbert, D. and Majury, D. (2006) 'Infertility and the Parameters of Discrimination Discourse', in D. Pothier and R. Devlin (eds) *Critical Disability Theory. Essays in Philosophy, Politics, Policy, and Law*, Vancouver: The University of British Columbia Press.
Glassner, B. (1989) 'Fitness and the postmodern self', *Journal of Health and Social Behavior*, 30: 180–91.
Goggin, G. and Newell, C. (2005) 'Foucault on the Phone. Disability and the Mobility of Government', in S. Tremain (ed) *Foucault and the Government of Disability*, Ann Arbor: The University of Michigan Press.
____. (2009) 'Fame and Disability: Christopher Reeve, Super Crips, and Infamous Celebrity', in R. M. Baird, S. E. Rosenbaum and S. K. Toombs (eds) *Disability: The Social, Political, and Ethical Debate*, Amherst: Prometheus Books.
Golan, T. (1998) 'The authority of shadows: the legal embrace of the X-ray', *Historical Reflections/Réflexions Historiques*, 24: 437–58.
____. (2004) 'The emergence of the silent witness: the legal and medical reception of X-rays in the USA,' *Social Studies of Science*, 34: 469–99.
Goldberg, E. (2001) *The Executive Brain: Frontal Lobes and the Civilized Mind*, Oxford: Oxford University Press.

Goldman, R., Klatz, R. and Berger, L. (1991) *Brain Fitness. Anti-aging Strategies for Achieving Super Mind Power*, New York: Doubleday.

Goodall, J. (2000) 'An Order of Pure Decision: Un-Natural Selection in the Work of Stelarc and Orlan', in M. Featherstone (ed) *Body Modification*, London: Sage Publications.

Gori, R. and Del Volgo, M. (2009) *La santé totalitaire. Essai sur la medicalisation de l'existence*, Paris: Flammarion.

Gray, C. H. (2002) 'In Defence of Prefigurative Art: The Aesthetics and Ethics of Orlan and Stelarc', in J. Zylinska (ed) *The Cyborg Experiments: The Extensions of the Body in the Media Age*, London and New York: Continuum.

Greco, M. (1993) 'Psychosomatic subjects and the "duty to be well": personal agency within medical rationality', *Economy and Society*, 22: 357–72.

Greenhalgh, T. and Wessely, S. (2004) '"Health for me": a sociocultural analysis of healthism in the middle classes', *British Medical Bulletin*, 69: 197–213.

Grmek, M. D. and Bernabeo, R. (1997) 'La machine du corps', in M. D. Grmek (ed) *Histoire de la pensée medicale en occident, 2. De la Renaissance aux Lumières*, Paris: Seuil.

Groce, N. E. (1999) *Everyone Here Spoke Sign Language. Hereditary Deafness in Martha's Vineyard*, Cambridge, Mass: Harvard University Press.

Grosz, E. A. (1989) *Sexual Subversions: Three French Feminists*, Sydney: Allen & Unwin.

____. (1990) 'The Body of Signification', in J. Fletcher and A. Benjamin (eds) *Abjection, Melancholia and Love: The Work of Julia Kristeva*, New York: Routledge.

____. (1994) *Volatile Bodies: Toward a Corporeal Feminism*, Bloomington: Indiana University Press.

____. (2004) *The Nick of Time: Politics, Evolution, and the Untimely*, Durham: Duke University Press.

Grove, A. W. (1997) 'Röntgen's ghosts: photography, X-rays, and the Victorian imagination', *Literature and Medicine*, 16: 141–73.

Gugerli, D. (1999) 'Soziotechnische Evidenzen. "Der pictorial turn" als Chance für die Geschichtswissenschaft', *Traverse* (Wissenschaft, die Bilder schafft), 3: 131–159.

Guthman, J. (2009) 'Neoliberalism and the Constitution of Contemporary Bodies', in E. Rothblum and S. Solovay (eds) *The Fat Studies Reader*, New York and London: New York University Press.

Hacking, I. (1995) 'The Looping Effects of Human Kinds', in D. Sperber, D. Premack and A. James-Premack (eds) *Causal Cognition: A Multidisciplinary Approach*, Oxford: Clarendon Press.

____. (1999) *The Social Construction of What?* Cambridge, Mass and London: Harvard University Press.

____. (2002) 'Making up people', in *Historical Ontology*, Cambridge, Mass: Harvard University Press.

Halberstam, J. and Livingston, I. (1995) *Posthuman Bodies*, Bloomington: Indiana University Press.

Halewood, P. (1996) 'Law's bodies: disembodiment and the structure of liberal property rights', *Iowa Law Review*, 18: 1331–93.

Hall, S. S. (2003) 'Is Buddhism Good for Your Health?', *New York Times*, available at http://www.nytimes.com/2003/09/14/magazine/is-buddhism-good-for-your-health.html?pagewanted=all&src=pm (accessed 30 September 2012).

Harcourt, G. (1987) 'Andreas Vesalius and the anatomy of antique sculpture', *Representations*, 17: 28–61.

Harding, J. (1997) 'Bodies at Risk. Sex, Surveillance and Hormone Replacement Therapy', in A. Petersen and R. Bunton (eds) *Foucault, Health and Medicine*, London: Routledge.

Harpham, G. (1987) *The Ascetic Imperative in Culture and Criticism*, Chicago and London: University of Chicago Press.

Harvey, D. (1990) *The Condition of Postmodernity. An Enquiry into the Origins of Cultural Change*, Oxford: Blackwell.

Hayles, N. K. (1999) *How We Became Posthuman: Virtual Bodies in Cybernetics, Literature, and Informatics*, Chicago: University of Chicago Press.

____. (2005) *My Mother Was a Computer: Digital Subjects and Literary Texts*, Chicago: University of Chicago Press.

Häyry, M. (2004) 'There is a difference between selecting a deaf embryo and deafening a hearing child', *Journal of Medical Ethics*, 30: 510–12.

Hayward, P. A. and Howard-Johnston, J. (eds) (2000) *The Cult of Saints in Late Antiquity and the Middle Ages: Essays on the Contribution of Peter Brown*, Oxford: Oxford University Press.

Heath, D., Rapp, R. and Taussig, K. (2004) 'Genetic Citizenship', in D. Nugent and J. Vincent (eds) *A Companion to the Anthropology of Politics*, Malden: Blackwell.

Hekman, S. (2008) 'Constructing the Ballast: An Ontology for Feminism', in S. Alaimo and S. Hekman (eds) *Material Feminisms*, Bloomington: Indiana University Press.

Henderson, J. et al (2009) '"Health is the number one thing we go for": Healthism, citizenship and food choice', *The Future of Sociology*, available at http://www.tasa.org.au/tasa-conference/past-tasa-conferences/2009-tasa-conference/ (accessed 23 March 2010).

Herring, J. and Chau, P. (2007) 'My body, your body, our bodies', *Medical Law Review*, 15: 34–61.

Hess, D. et al (2008) 'Science, Technology, and Social Movements', in M Lynch and J. Wajcman (eds) *The Handbook of Science and Technology Studies*, 3rd edn, Cambridge, Mass: MIT Press.

Hesse-Biber, S. (1997) *Am I Thin Enough Yet?: The Cult of Thinness and the Commercialization of Identity*, New York and Oxford: Oxford University Press.

Heyes, C. J. (2002) 'Back to the Rough Ground: Wittgenstein, Essentialism and Feminist Methods', in N. Scheman (ed) *Re-reading the Canon: Feminist Interpretation of Ludwig Wittgenstein*, University Park: Pennsylvania State University Press.

____. (2007) *Self-Transformations. Foucault, Ethics, and Normalized Bodies*, New York: Oxford University Press.

Hillman, D. (1996) 'Hamlet, Nietzsche, and Visceral Knowledge', in M. O'Donovan-Anderson (ed) *The Incorporated Self. Interdisciplinary Perspectives on Embodiment*, Lanham: Rowman and Littlefield Publishers.

____. (1997) 'Visceral Knowledge. Shakespeare, Skepticism, and the Interior of the Early Modern Body', in D. Hillman and C. Mazzio (eds) *The Body in Parts. Fantasies of Corporeality in Early Modern Europe*, New York and London: Routledge.

Hippocrates. (1983) *Hippocratic Writings*, trans. J. Chadwick and W. N. Mann, London and New York: Penguin.

Hirschauer, S. (2006) 'Animated corpses: communicating with post mortals in an anatomical exhibition', *Body & Society*, 12: 25–52.

Hoffman, P. (2001) 'Introduction II. How Todes Rescues Phenomenology from the Threat of Idealism', in S. Todes (ed), *Body and World*, Cambridge, Mass: MIT Press.

Holland, S. (2001) 'Contested commodities at both ends of life: buying and selling gametes, embryos, and body tissues', *Kennedy Institute of Ethics Journal*, 11: 263–84.

Holmes, B. (2010) *The Symptom and the Subject. The Emergence of the Physical Body in Ancient Greece*, Princeton and Oxford: Princeton University Press.

Honneth, A. (1993) *The Critique of Power. Reflective Stages in a Critical Social Theory*, trans. Kenneth Baynes, Cambridge, Mass and London: MIT Press.

Horn, C. (1998) *Antike Lebenskunst: Glück und Moral von Sokrates bis zu den Neuplatonikern*, München: Beck.

Houis, J., Mieli, P. and Stafford, M. (eds) (1999) *Being Human. The Technological Extensions of the Body*, New York: Après-Coup.

Howson, A. (2005) *Embodying Gender*, London: Sage Publications.

____. and Inglis, D. (2001) 'The body in sociology: tensions inside and outside sociological thought', *Sociological Review*, 49: 297–317.

Hoy, D. C. (1999) 'Critical Resistance: Foucault and Bourdieu', in G. Weiss and H. Haber (eds) *Perspectives on Embodiment. The Intersections of Nature and Culture*, London and New York: Routledge.

Hubbard, R. (2010) 'Abortion and Disability: Who Should and Should not Inhabit the World?', in L. J. Davis (ed) *The Disability Studies Reader*, 3rd edn, New York: Routledge.

Hughes, B. (2008) 'What can a Foucauldian Analysis Contribute to Disability Theory?', in Shelley Tremain (ed) *Foucault and the Government of Disability*, Ann Arbor: University of Michigan Press.

____. (2009) 'Disability activisms: social model stalwarts and biological citizens', *Disability & Society*, 24: 677–78.

Husserl, E. (1952) *Ideen zu einer Reinen Phänomenologie und phänomenologischen Philosophie. Zweites Buch: Phänomenologische Untersuchungen zur Konstitution*, (Husserliana, Band IV), Den Haag: Martinus Nijhoff.

Hyde, A. (1997) *Bodies of Law*, Princeton: Princeton University Press.

Hyde, M. and Power, D. (2006) 'Some ethical dimensions of cochlear implantation for deaf children and their families', *Journal of Deaf Studies and Deaf Education*, 11: 102–11.

Illich, I. (2010) [1975] *Limits to Medicine* (originally published as: *Medical Nemesis: The Expropriation of Health*), London: Marion Boyards.

Jackman, M. (2007) 'Health care and equality: is there a cure?', *Health Law Journal*, 15: 87–142.

Jackson, J. (2001) 'Chronic pain and the tensions between the body as subject and object', in T. J. Csordas (ed) *Embodiment and Experience. The Existential Ground of Culture and Self*, Cambridge: Cambridge University Press.

James, W. (1952) [1890] *The Principles of Psychology* [*Great Books of the Western World*], Chicago: William Benton, Encylopaedia Britannica, Inc.

____. (1968) [1904] 'The Experience of Activity' ('Essays in Radical Empiricism'), in J. Mcdermott (ed) *The Writings of William James*, New York: The Modern Library.

Jeffreys, S. (2005) *Beauty and Misogyny: Harmful Cultural Practices in the West*, London: Routledge.

Jeggle, U. (1980) 'Im Schatten des Körpers. Vorüberlegungen zu einer Volkskunde der Körperlichkeit', *Volkskunde*, 76: 169–88.

Jenkins, J. H. (ed) (2011) *Pharmaceutical Self: The Global Shaping of Experience in an Age of Psychopharmacology*, Santa Fe: School for Advanced Research Press.

Joas, H. (1996) *The Creativity of Action*, trans. Jeremy Gaines and Paul Keast, Chicago: University of Chicago Press.

Johnson, M. (1987) *The Body in the Mind: The Bodily Basis of Meaning, Imagination, and Reason*, Chicago: University of Chicago Press.

____. (1999) 'Embodied Reason', in G. Weiss and H. F. Haber (eds) *Perspectives on Embodiment. The Intersections of Nature and Culture*, London and New York: Routledge.

Jonas, H. (2001) *The Phenomenon of Life. Toward a Philosophical Biology*, Evanston, Il: Nortwestern University Press.

Jones, M. and Sofia, Z. (2002) 'Stelarc and Orlan in the Middle Ages', in J. Zylinska (ed) *The Cyborg Experiments: The Extensions of the Body in the Media Age*, London and New York: Continuum.

Joyce, K. A. (2005) 'Appealing images magnetic resonance imaging and the production of authoritative knowledge', *Social Studies of Science*, 35: 437–62.

____. (2008) *Magnetic Appeal. MRI and the Myth of Transparency*, Ithaca: Cornell University Press.

Jütte, R. (2000) *Geschichte der Sinne. Von der Antike bis zum Cyberspace*, München: Verlag C. H. Beck.

Kahn, J. (2010) 'Surrogate Markers and Surrogate Marketing in Biomedicine. The Regulatory Etiology and Commercial Progression of "Ethnic" Drug Development', in A. Clarke, L. Mamo, J. R. Foster, J. R. Fishman and Janet K. Shim (eds) *Biomedicalization. Technoscience, Health and Illness in the U.S.*, Durham and London: Duke University Press.

Keith, A. and Vaage, L. E. (1999) 'Imperial Asceticism: Discipline of Domination', in L. E. Vaage and V. A. Wimbush (eds) *Asceticism in the New Testament: Redefining the Interpretive Practice of a Discipline*, London and New York: Routledge.

Kemp, M (1998) 'A Perfect and Faithful Record': Mind and Body in Medical Photography before 1900', in A. Thomas (ed) *Beauty of Another Order. Photography in Science*, New Haven and London: Yale University Press.

Kermit, P. (2009) 'Cochlear Implants, Linguistic Rights and "Open Future" Arguments', in K. Kristianse, S. Vehmans and T. Shakespeare (eds) *Arguing About Disabilities. Philosophical Perspectives*, New York: Routledge.

Kevles, B. H. (1997) *Naked to the Bone. Medical Imaging in the Twentieth Century*, Reading, Mass: Helix Books.

Key, L. E. (1996) 'Voluntary disabilities and the ADA: a reasonable interpretation of "reasonable accommodations"', *Hastings Law Journal*, 48: 75–104.

Kimbrell, A. (1993) *The Human Body Shop: The Engineering and Marketing of Life*, New York: HarperCollins Publishers.

King, L. S. and Meehan, M. C. (1973) 'A history of the autopsy', *The American Journal of Pathology*, 73, 514–44.

Kirkland, A. (2008) *Fat Rights. Dilemmas of Difference and Personhood*, New York and London: New York University Press.

Kittay, E. (1998) *Love's Labor: Essays on Women, Equality, and Dependence*, New York: Routledge.

____. (2001) 'When caring is just and justice is caring: justice and mental retardation', *Public Culture*, 13: 557–79.

Klestinec, C. (2004) 'A history of anatomy theaters in sixteenth century Padua', *Journal of the History of Medicine and Allied Sciences*, 59: 375–412.

____. (2007) 'Civility, comportment, and the anatomy theater: Girolamo Fabrici and his medical students in Renaissance Padua', *Renaissance Quarterly*, 60: 434–63.

____(2011) *Theaters of Anatomy: Students, Teachers, and Traditions of Dissection in Renaissance Venice*, Baltimore: Johns Hopkins University Press.

Kristeva, J. (1982) *Powers of Horror: An Essay on Abjection*, trans. L. S. Roudiez, New York: Columbia University Press.

Kroll-Smith, S. (2003) 'Popular media and "excessive daytime sleepiness": a study of rhetorical authority in medical sociology', *Sociology of Health and Illness*, 25: 625–43.

Krüger-Fürhoff, I. M. (2001) *Der versehrte Körper. Revisionen des Klassizistischen Schönheitsideals*, Göttingen: Wallstein Verlag.

Kurzweil, R. (2000) *The Age of Spiritual Machines. When Computers Exceed Human Intelligence*, New York and London: Penguin.

Lacan, J. (ed) (2001) 'The Mirror Stage as Formative of the Function of the I as Revealed in Psychoanalytic Experience', in *Écrits. A Selection*, trans. A. Sheridan, London: Routledge.

Lacoue-Labarthe, P. and Nancy, J. L. (1988) *The Literary Absolute. The Theory of Literature of German Romanticism*, trans. Philip Barnard and Cheryl Lester, Albany: State University of New York Press.

Lakoff, G. (1990) *Women, Fire, and Dangerous Things. What Categories Reveal About the Mind*, Chicago and London: University of Chicago Press.

____. and Johnson, M. (1980) *Metaphors We Live By*, Chicago: University of Chicago Press.

____. and Johnson, M. (1999) *Philosophy in the Flesh: The Embodied Mind and Its Challenge to Western Thought*, New York: HarperCollins Publishers.

Landow, G. P. (1997) *Hypertext 2.0: The Convergence of Contemporary Critical Theory and Technology*, Baltimore: Johns Hopkins University Press.

Landweer, H. (2002) 'Konstruktion und begrenzte Verfügbarkeit des Körpers', in A. Barkhaus and A. Fleig (eds) *Grenzverläufe. Der Körper als Schnitt-Stelle*, München: Wilhelm Fink Verlag.

Lane, H. and Grodin, M. (1997) 'Ethical issues in cochlear implant surgery: an exploration into disease, disability, and the best interests of the child', *Kennedy Institute of Ethics Journal*, 7: 231–51.

Lantos, J. (ed) (2011) *Controversial Bodies. Thoughts on the Public Display of Plastinated Corpses*, Baltimore: Johns Hopkins University Press.

Latour, B. (2003) 'The Promises of Constructivism', in D. Ihde and E. Selinger (ed). *Chasing Technology: Matrix of Materiality*, Bloomington, Indiana: Indiana University Press.

____. (2004) 'Why has critique run out of steam? From matters of fact to matters of concern', *Critical Inquiry*, 30: 225–48.

Lauretis, T. de (2003) 'Becoming inorganic (David Cronenberg's eXistenZ and Freud's theory of drives)', *Critical Inquiry*, 29: 547–70.

LEAF (2004) 'The Supreme Court of Canada decides that autistic children do not have a right to health services', available at http://www.leaf.ca/legal-media.html#November1 (accessed 19 November 2004).

Lebesco, K. (2010) 'Fat Panic and the New Morality', in J. M. Metzl and A. Kirkland (eds) *Against Health. How Health Became the New Morality*, New York and London: New York University Press.

Le Breton, D. (1998), 'Réflexions sur la médicalisation de la douleur', in P. Aïach and D. Delanoë (eds) *L'ère de la médicalisation. Ecce homo sanitas*, Paris: Anthropos.

____. (1999) *L'adieu au corps*, Paris: Éditions Métailié.

____. (2000) *La sociologie du corps*, 4th edn, Paris: Presses Universitaires de France.

____. (2002a) *Signes d'identité. Tatouages, piercings et autres marques corporelles*, Paris: Editions Métailié.

____. (2002b) *Conduites à risque: des jeux de mort au jeu de vivre*, Paris: PUF.

____. (2003) *Anthropologie du corps et modernité*, Paris: PUF.

Leder, D. (1990) *The Absent Body*, Chicago and London: University of Chicago Press.

____. (1992) 'A Tale of Two Bodies: The Cartesian Corpse and the Lived Body', in D. Leder (ed) *The Body in Medical Thought and Practice*, Dordrecht, Boston and London: Kluwer Academic Publishers.

Leibing, A. and Tournay, V. (eds) (2010) *Les technologies de l'espoir. La Fabrique d'une histoire à accomplir*, Quebec: Presses Universitaires de Laval.

Leigh, P. R. (2003) 'Interest convergence and desegregation in the Ohio Valley', *Journal of Negro Education*, 72: 269–96.

Levin, J. (2008) Bodies and subjects in Merleau-Ponty and Foucault: towards a phenomenological/poststructuralist feminist theory of embodied subjectivity', Unpublished Thesis. The Pennsylvania State University – The Graduate School, College of the Liberal Arts.

Levy, N. (2002) 'Deafness, culture, and choice', *Journal of Medical Ethics*, 28: 284–85.

Lindt, G. (2002) 'Asceticism in Sociological Perspective', in V. L. Wimbush and R. Valantasis (eds) *Asceticism*, New York: Oxford University Press.

Linke, U. (2005) 'Touching the corpse: the unmaking of memory in the body museum', *Anthropology Today*, 21: 13–19.

Linton, S. (2010) 'Reassigning Meaning', in L. J. Davis (ed) *The Disability Studies Reader*, 3rd edn, New York and London: Routledge.

List, E. (1994) 'Wissende Körper – Wissenskörper – Maschinenkörper. Zur Semiotik der Leiblichkeit', *Die Philosophin. Forum für Feministische Theorie und Philosophie*, 10: 9–26.

Lizama, N. (2009) 'The Post-biological Body: Horror, Nostalgia, and the Visible Human Project', in E. Klaver (ed) *The Body in Medical Culture*, Albany: State University of New York Press.

Lock, M. (1993) *Encounters with Aging: Mythologies of Menopause in Japan and North America*, Berkeley: University of California Press.

____. and Nguyen, V. (2010) *An Anthropology of Biomedicine*, Malden and Oxford: Wiley-Blackwell.

Lohse, B. (1969) *Askese und Mönchtum in der Antike und in der Alten Kirche*, München: R. Oldenbourg.

Lorenz, M. (2000) *Leibhaftige Vergangenheit: Einführung in die Körpergeschichte*, Tübingen: Edition diskord.

Lunsingh Scheurleer, T. H. (1975) 'Un amphithéâtre d'anatomie moralisée', in T. H. Lunsingh Scheurleer and G. H. M. Posthumus Meyjes (eds) *Leiden University in the Seventeenth Century: An Exchange of Learning*, Leiden: Brill.

Lupton, D. (1994) *Medicine as Culture. Illness, Disease and the Body in Western Societies*, London: Sage Publications.

____. (1995) *The Imperative of Health: Public Health and the Regulated Body*, London: Sage Publications.

____. (2003) *Medicine as Culture. Illness, Disease and the Body in Western Societies*, 2nd edn, London: Sage Publications.

Lyon, M. L. and Barbalet, J. M. (2001) 'Society's body: emotion and the "somatization" of social theory', in T. J. Csordas (ed) *Embodiment and Experience. The Existential Ground of Culture and Self*, Cambridge: Cambridge University Press.

Macleod, R. B. and Pick Jr, H. (eds) (1974) *Perception. Essays in Honor of James J. Gibson*, Ithaca and London: Cornell University Press.

Macpherson, B. C. (1962) *The Political Theory of Possessive Individualism: Hobbes to Locke*, Oxford: Oxford University Press.

Malabou, C. (2008) *What Should We Do With Our Brain?* trans. S. Rand, New York: Fordham University Press.

Malhotra, R. A. (2006) 'Justice as Fairness in Accommodating Workers with Disabilities and Critical Theory: The Limitations of a Rawlsian Framework for Empowering People with Disabilities in Canada', in D. Pothier and R. Devlin (eds) *Critical Disability Theory. Essays in Philosophy, Politics, Policy, and Law*, Vancouver: The University of British Columbia Press.

——. and Hansen, R. (2011) 'The United Nations Convention on the rights of persons with disabilities and its implications for the equality rights of Canadians with disabilities: the case of education', *Windsor Yearbook of Access to Justice*, 29: 1.

Mandressi, R. (2003) *Le regard de l'anatomiste. Dissections et invention du corps en Occident*, Paris: Seuil.

Mann, T. (1999) *The Magic Mountain*, trans. H. T. Lowe-Porter, London: Vintage.

Mark, V. H., and Mark, J. P. (1991) *Brain Power. A Neurosurgeon's Complete Program to Maintain and Enhance Brain Fitness Throughout Your Life*, Boston: Houghton Mifflin.

Markus, R. A. (1998) *The End of Ancient Christianity*, Cambridge: Cambridge University Press.

Marshall, T. H. (1950) *Citizenship and Social Class and Other Essays*, Cambridge: Cambridge University Press.

Martin, E. (1987). *The Woman in the Body: A Cultural Analysis of Reproduction*. Boston: Beacon Press.

____. (2010) 'Self-making and the brain', *Subjectivity*, 3: 366–81.

Martz, E. (2001) 'Acceptance of imperfection', *Disability Studies Quarterly*, 21: 160–65.

Mazzio, C. (1997) 'Introduction: individual parts', in D. Hillman and C. Mazzio (eds) *The Body in Parts. Fantasies of Corporeality in Early Modern Europe*, New York and London: Routledge.

McAuley, E. et al. (2000) 'Social relations, physical activity and well-being in older adults', *Preventive Medicine*, 31: 608–617.

McDonagh, E. (2009) *The Motherless State: Women's Political Leadership and American Democracy*, Chicago and London: University of Chicago Press.

McFarlane, A. (1992) 'On individualism', *Proceedings of the British Academy*, 82: 171–99.

McGushin, E. (2007) *Foucault's Askesis: An Introduction to the Philosophical Life*, Evanston, Il: Northwestern University Press.

McGushin, E. F. (2007) *Foucault's Askesis: An Introduction to the Philosophical Life*, Evanston, Il: Northwestern University Press.

McLaren, M. A. (2002) *Feminism, Foucault and Embodied Subjectivity*, Albany: State University of New York Press.

McLellan, F. (2002) 'Controversy over deliberate conception of deaf child', *The Lancet*, 359: 1315.

McMichael, A. J. and Beaglehole, R. (2000) 'The changing global context of public health', *The Lancet*, 356: 495–99.

McNay, L. (1991) 'The Foucauldian body and the exclusion of experience', *Hypatia*, 6: 125–39.

Merleau-Ponty, M. (1964) *The Primacy of Perception: And Other Essays on Phenomenological Psychology, the Philosophy of Art, History And Politics*, trans. William Cobb, Evanston, Il: Northwestern University Press.

____. (2002) *Phenomenology of Perception*, trans. Colin Smith, London and New York: Routledge.

Miah, A. and Rich, E. (2008) *The Medicalization of Cyberspace*, London and New York: Routledge.

Miller, P. and Rose, N. (2008) *Governing the Present. Administering Economic, Social and Personal Life*, Cambridge and Malden: Polity Press.

Milner, H. R. (2008) 'Critical race theory and interest convergence as analytic tools in teacher education policies and practices', *Journal of Teacher Education*, 59: 332–46.

Minsky, M. (1988) *The Society of Mind*, New York: Simon and Schuster.

Molsberry, R. (2009) 'More than an Inspiration', in R. M. Baird, S. E. Rosenbaum and S. K. Toombs (eds) *Disability: The Social, Political, and Ethical Debate*, Amherst: Prometheus Books.

Mor, S. (2006) 'Between charity, welfare, and warfare: a disability legal studies analysis of privilege and neglect in Israeli disability policy', *Yale Journal of Law and the Humanities*, 18: 63–136.

Moraes, E. R. (2002) *O corpo impossível: A decomposição da figura humana de Lautréamont a Bataille*, São Paulo: Iluminuras.

Morris, D. (1993) *The Culture of Pain*, Berkeley: University of California Press.

____. (2000) *Illness and Culture in the Postmodern Age*, Berkeley and Los Angeles: University of California Press.

Moscoso, J. (2012) *Pain: A Cultural History*, trans. S. Thomas and P. House, Basingstoke and New York: Palgrave Macmillan.

Moss, P. and Teghtsoonian, K. (eds) (2008) *Contesting Illness: Processes and Practices*, Toronto: University of Toronto Press.

Mundy L. (2002) 'A world of their own', *Washington Post*, 31: W22.

Myers, J. (1992) 'Nonmainstream body modification', *Journal of Contemporary Ethnography*, 21: 267–306.

Nadesan, M. H. (2005) *Constructing Autism: Unravelling the 'Truth' and Understanding the Social*, London and New York: Routledge.

——. (2008) *Governmentality, Biopower, and Everyday Life*, London and New York: Routledge.

Negrin, L. (2008) *Appearance and Identity: Fashioning the Body in Postmodernity*, New York: Palgrave Macmillan.

Nelson, A. (2004) 'Declaration from the autism community that they are a minority group', available at http://www.prweb.com/releases/2004/11/prweb179444.htm (accessed 13 August 2008).

Netherland, J. (2011) '"We Haven't Sliced Open Anyone's Brain Yet": Neuroscience, Embodiment and the Governance of Addiction', in M. Pickersgill, I. V. Keulen (eds) *Sociological Reflections on the Neurosciences* (Advances in Medical Sociology, volume 13), Bingley: Emerald Group Publishing Limited.

Nettleton, S. (1997) 'Governing the Risky Self: How to Become Healthy, Wealthy and Wise', in A. Petersen and R. Bunton (eds) *Foucault, Health and Medicine*, London: Routledge.

____. (2004) 'The emergence of e-scaped medicine?' *Sociology*, 38: 661–79.

Nguyen, V. (2005) 'Antiretroviral Globalism, Biopolitics, and Therapeutic Citizenship', in A. Ong and S. J. Collier (eds) *Global Assemblages: Technology, Politics, and Ethics as Anthropological Problems*, Malden: Blackwell.

Nissenbaum, S. (1980) *Sex, Diet, and Debility in Jacksonian America. Sylvester Graham and Health Reform*, Westport, Connecticut: Greenwood.

Novas, C. (2006) 'The political economy of hope: patients' organizations, science and biovalue', *BioSocieties*, 1: 289–305.

Novas, C. and Rose, N. (2000) 'Genetic risk and the birth of the somatic individual', *Economy and Society*, 29: 485–513.

Nussbaum, M. (2006) *Frontiers of Justice: Disability, Nationality, Species Membership*, Cambridge, Mass: Harvard University Press.

____. (2010) 'The Capabilities of People with Cognitive Disabilities', in E. F. Kittay and L. Carlson (eds) *Cognitive Disability and its Challenge to Moral Philosophy*, Malden and Oxford: Wiley-Blackwell.

O'Brien, R. (ed) (2004) *Voices From the Edge: Narratives about the Americans With Disabilities Act*, New York: Oxford University Press.

O'Bryan, J. C. (2005) *Carnal Art: Orlan's Refacing*, Minneapolis: University of Minnesota Press.

O'Connor, C. and Joffe, H. (2013) 'How has neuroscience affected lay understandings of personhood? A review of the evidence', *Public Understanding of Science*, 22(3): 254–268.

Oetjen, A. and Wacker, H. (1993) *Organischer Horror: Die Filme des David Cronenberg*, Meitingen: Corian-Verlag.

Offit, P. A. (2008) *Autism's False Prophets*, New York: Columbia University Press.

Oliver, M. (1990) *The Politics of Disablement*, London: Macmillan.

____. (1996) *Understanding Disability: From Theory to Practice*, Basingstoke: Macmillan.

Ong, A. (1999) *Flexible Citizenship. The Cultural Logic of Transnationality*, Durham and London: Duke University Press.

____. (2005) 'Ecologies of Expertise. Assembling Flows, Managing Citizenship', in A. Ong and S. J. Collier (eds) *Global Assemblages: Technology, Politics, and Ethics as Anthropological Problems*, Malden: Blackwell.

____. (2006a) 'Mutations in citizenship', *Theory, Culture, and Society*, 23: 499–505.

____. (2006b) *Neoliberalism as Exception. Mutations in Citizenship and Sovereignty*, Durham and London: Duke University Press.

Orbach, S. (2009) *Bodies*, New York: Picador.

Orlan. (1989) 'Carnal Art Manifesto' (*L'Art charnel*), available at http://orlan.eu/adriensina/manifeste/carnal.html (accessed 25 November 2012).

——. (2002) 'The Virtual and/or the Real', in J. Zylinska (ed) *The Cyborg Experiments: The Extensions of the Body in the Media Age*, London and New York: Continuum.

Orsini, M. (2008) 'Hepatitis C and the Dawn of Biological Citizenship: Unravelling the Policy Implications', in P. Moss and K. Teghtsoonian (eds) *Contesting Illness: Processes and Practices*, Toronto, Buffalo and London: University of Toronto Press.

____. (2009) 'Contesting the Autistic Subject: Biological Citizenship and the Autism/Autistic Movement', in S. Murray and D. Holmes (eds) *Critical Interventions in the Ethics of Health Care*, London: Ashgate Publishing.

____. and Smith, M. (2010) 'Social movements, knowledge and public policy: the case of autism activism in Canada and the US', *Critical Policy Studies*, 4: 38–57.

Ortega, F. (1997) *Michel Foucault: Rekonstruktion der Freundschaft*, Munich: Wilhelm Fink.

____. (2001) 'Michel Foucault: os sentidos da subjetividade', in B. Bezerra Jr and C. A. Plastino (eds) *Corpo, afeto, linguagem: a questão do sentido hoje*, Rio de Janeiro: Contracapa.

____. (2007) 'Self e continuidade: entre Winnicott e William James', in B. Bezerra Jr and F. Ortega (eds) *Winnicott e seus interlocutores*, Rio de Janeiro: Relume Dumará.

——. (2009a) 'The cerebral subject and the challenge of neurodiversity', *BioSocieties*, 4: 425–45.

——. (2009b) 'Deficiência, autismo e neurodiversidade', *Ciência e Saúde Coletiva*, 14: 67–77.

____. (2011) 'Toward a Genealogy of Neuroascesis', in F. Ortega and F. Vidal (eds) *Neurocultures: Glimpses into an Expanding Universe*, Frankfurt and New York: Peter Lang.

____. and Choudhury, S. (2011) "'Wired up differently': autism, adolescence and the politics of neurological identities', *Subjectivity*, 4: 323–45.

____. and Vidal, F. (2010) 'Les (dés) espoirs du cerveau: neuroascèse et neuroéthique', in A. Leibing and V. Tournay (eds) *Les technologies de l'espoir. La fabrique d'une histoire à accomplir*, Quebec: Presses Universitaires de Laval.

____. and Vidal, F. (eds) (2011) *Neurocultures: Glimpses into an Expanding Universe*, Frankfurt and New York: Peter Lang.

____. and Vidal, F. (2013) 'Brains in literature/literature in the brain', *Poetics Today*, 34(3): 327–360.

Outram, D. (1989) *The Body and the French Revolution. Sex, Class and Political Culture*, New Haven and London: Yale University Press.

Padden, C. and Humphries, T. (1988) *Deaf in America: Voices From a Culture*, Cambridge, Mass: Harvard University Press.

____. (2005) *Inside Deaf Culture*, Cambridge, Mass London: Harvard University Press.

____. (2010) 'Deaf People: A Different Center', in L. J. Davis (ed) *The Disability Studies Reader*, 3rd edn, New York: Routledge.

Pamboukian, S. (2001) "'Looking radiant": science, photography and the X-ray craze of 1896', *Victorian Review*, 27: 56–74.

Park, K. (1994) 'The criminal and the saintly body: autopsy and dissection in Renaissance Italy', *Renaissance Quarterly*, 47: 1–33.

____. (2010) *Secrets of Women. Gender, Generation, and the Origins of Human Dissection*, New York: Zone.

Parlette, S. (1997) *The Brain Workout Book*, New York: M. Evans and Company, Inc.

Parry, B. (2012) 'Economies of Bodily Commodification', in T. J. Barnes, J. Peck and E. Sheppard (eds) *The Wiley-Blackwell Companion to Economic Geography*, Chichester, UK: Wiley-Blackwell.

Peace, W. J. (2009) 'Wishing for Kryptonite: A Response to Christopher Reeve's Pursuit of Cure', in R. M. Baird, S. E. Rosenbaum and S. K. Toombs (eds) *Disability: the Social, Political, and Ethical Debate*, Amherst: Prometheus Books.

Petersen, A. (1997) 'Risk, Governance and the New Public Health', in A. Petersen and R. Bunton (eds) *Foucault, Health and Medicine*, London: Routledge.

____. (2005) 'Securing our genetic health: engendering trust in UK Biobank', *Sociology of Health & Illness*, 27: 271–92.

____. (2007) *The Body in Question: A Socio-Cultural Approach*, London: Routledge.

Petit, J. (2010) 'A Husserlian, Neurophenomenologic Approach to Embodiment', in D. Schmicking and S. Gallagher (eds) *Handbook of Phenomenology and Cognitive Science*, Dordrecht, New York, Heidelberg and London: Springer.

Petryna, A. (2002) *Life Exposed: Biological Citizens After Chernobyl*, Princeton: Princeton University Press.

Pickersgill, M., Cunningham-Burley, S. and Martin, P. (2011) 'Constituting neurologic subjects: neuroscience, subjectivity and the mundane significance of the brain', *Subjectivity*, 4: 346–65.

Pitts, V. (2000) 'Body Modification, Self-mutilation and Agency in Media Accounts of a Subculture', in M. Featherstone (ed) *Body Modification*, London: Sage Publications.

____. (2003) *In the Flesh. The Cultural Politics of Body Modification*, Basingstoke and New York: Palgrave.

Pitts-Taylor, V. (2010) 'The plastic brain: neoliberalism and the neuronal self', *Health*, 14: 635–52.

Plant S. (1997) *Zeroes and ones: Digital Women and the New Technoculture*, London: Fourth Estate.

Plato (2005) *The Republic*, in E. Hamilton and H. Cairns (eds) *The Collected Dialogues of Plato including The Letters*, 7th edn, Princeton: Princeton University Press.

Porter, J. (2005) 'Foucault's ascetic ancients', *Phoenix*, 59 (1/2): 121–32.

Porter, R. (1992) 'History of the Body', in P. Burke (ed) *New Perspectives in Historical Writing*, University Park: Pennsylvania State University Press.

____. (1993) 'The Rise of Physical Examination', in W. F. Bynum and R. Porter (eds) *Medicine and the Five Senses*, Cambridge: Cambridge University Press.

____. (1999) *The Greatest Benefit to Humankind: A Medical History of Humanity*, New York: W. W. Norton & Company, Inc.

____. (2004) *Blood and Guts: A Short History of Medicine*, New York: W. W. Norton & Company, Inc.

Poster, M. (2002) 'High-Tech Frankenstein, or Heidegger Meets Stelarc', in J. Zylinska (ed) *The Cyborg Experiments: The Extensions of the Body in the Media Age*, London and New York: Continuum.

Pothier, D. (2006) 'Appendix: Legal Developments in the Supreme Court of Canada Regarding Disability', in D. Pothier and R. Devlin (eds) *Critical Disability Theory. Essays in Philosophy, Politics, Policy, and Law*, Vancouver: The University of British Columbia Press.
——. and Devlin, R. (eds) (2006) *Critical Disability Theory. Essays in Philosophy, Politics, Policy, and Law*, Vancouver: The University of British Columbia Press.
Pouchelle, M. C. (1983) *Corps et chirurgie a l'apogée du moyen age*, Paris: Flammarion.
Power, D. (2005) 'Models of deafness: cochlear implants in the Australian daily press', *Journal of Deaf Studies and Deaf Education*, 10: 451–59.
Prentice, R. (2005) 'The anatomy of a surgical simulation: the mutual articulation of bodies in and through the machine,' *Social Studies of Science*, 35: 837–66.
Prince, M. J. (2008) 'Claiming a Disability Benefit as Contesting Social Citizenship', in P. Moss and K. Teghtsoonian (eds) *Contesting Illness: Processes and Practices*, Toronto and Buffalo and London: University of Toronto Press.
Rabinow, P. (1996a), 'Severing the Ties: Fragmentation and Dignity in Late Modernity', in *Essays on the Anthropology of Reason*, Princeton: Princeton University Press.
____. (1996b), 'Artificiality and Enlightenment: From Sociobiology to Biosociality', in *Essays on the Anthropology of Reason*, Princeton: Princeton University Press.
____. (2005) 'Midst Anthropology's Problems', in A. Ong and S. J. Collier (eds) *Global Assemblages: Technology, Politics, and Ethics as Anthropological Problems*, Malden: Blackwell.
—— (2008) 'Afterwork: concept work', in S. Gibbon and C. Novas (eds) *Biosocialities, Genetics and the Social Sciences. Making Biologies and Identities*, London and New York: Routledge.
Race, K. (2009a) *Pleasure Consuming Medicine: The Queer Politics of Drugs*, Durham: Duke University Press.
____. (2009b) 'Queer Substances: Of Drugs, Dogs and Other Piggy Practices', in N. Sullivan and S. Murray (eds) *Somatechnics: Queering the Technologisation of Bodies*, Aldershot: Ashgate.
Ramos, A. (1994) 'Technologies of the self: truth, asceticism, and autonomy', *Bulletin de la Société Américaine de Philosophie de Langue Française*, 6: 20–29.
Rapp, C. (1999) '"For the Next to God You are My Salvation": Reflections on the Rise of the Holy Man in Late Antiquity' in J. Howard-Johnston and P. A. Hayward (eds) *The Cult of Saints in Late Antiquity and the Middle Ages. Essays on the Contribution of Peter Brown*, Oxford: Oxford University Press.
Rapp, R. (1999) *Testing Women, Testing the Fetus: The Social Impact of Amniocentesis in America*, New York: Routledge.
Rawls, J. (1971) *A Theory of Justice*, Cambridge, Mass: Harvard University Press.
____. (1993) *Political Liberalism*, New York: Columbia University Press.
Rees, D., and Rose, S. (eds) (2004) *The New Brain Sciences: Perils and Prospects*, Cambridge: Cambridge University Press.
Rees, T. (2010) 'Being neurologically human today: life and science and adult cerebral plasticity (an ethical analysis)', *American Ethnologist*, 37: 150–66.
Reihling, H. P. (2010) 'Rejeitar ou priorizar a vida? Ambiguidades da biopolítica da Aids no Uruguai', *Ciência e saúde coletiva*, 15: 1159–68.
Reiser, S. J. (1990) *Medicine and the Reign of Technology*, Cambridge: Cambridge University Press.

Rheinberger, H. J. (1997) *Toward a History of Epistemic Things: Synthesizing Proteins in the Test Tube*, Stanford: Stanford University Press.
Richardson, H. (2006) 'Rawlsian social-contract theory and the severely disabled', *Ethics*, 10: 419–62.
Riches, S. (ed) (2012) *The Philosophy of David Cronenberg*, Lexington: The University Press of Kentucky.
Richlin, A. (1998) 'Foucault's History of Sexuality: A Useful Theory for Women?', in D. H. J. Larmour, P. A. Miller and C. Platter (eds) *Rethinking Sexuality: Foucault and Classical Antiquity*, Princeton: Princeton University Press.
Riepe, M. (1996) 'Karzinome der Lust: Körper und Fremdkörper in David Cronenbergs Filmen', *Kunstforum*, 133: 200–207.
____. (2002) *Bildgeschwüre: Körper und Fremdkörper im Kino David Cronenbergs*, Bielefeld: Transcript.
Rioux, M. H. and Valentine, F. (2006) 'Does Theory Matter? Exploring the Nexus between Disability, Human Rights, and Public Policy', in D. Pothier and R. Devlin (eds) *Critical Disability Theory. Essays in Philosophy, Politics, Policy, and Law*, Vancouver: The University of British Columbia Press.
Roberts, D. (2010) 'The Social Immorality of Health in the Gene Age: Race, Disability, and Inequality', in J. M. Metzl and A Kirkland (eds) *Against Health: How Health Became the New Morality*, New York and London: New York University Press.
Roberts, E. F. S. (2008) 'Biology, Sociality and Reproductive Modernity in Ecuadorian *In-vitro* Fertilization: the Particulars of Place', in S. Gibbon and C. Novas (eds) *Biosocialities, Genetics and the Social Science:. Making Biological Identities,* London and New York: Routledge.
Roche, D. (2003) 'David Cronenberg's having to make the word be flesh', *Post Script. Essays in Film and the Humanities*, 23: 72–87.
Rodis-Lewis, G. (1995) *Descartes*, Paris: Calmann-Levy.
Rodley, C. (1997) *Cronenberg on Cronenberg*, London: Faber and Faber.
Roelens, N. and Strauven, W. (eds) (2001) *Homo Orthopedicus. Le corps et ses prothèses à l'époque (post)moderniste*, Paris : L'Harmattan.
Roepstorff, A. (2007) 'Navigating the Brainscape: When Knowing Becomes Seeing', in C. Grasseni (ed) *Skilled Visions: Between Apprenticeship and Standards*, Oxford: Berghahn Books.
Rose, N. (1988) 'Calculable minds and manageable individuals', *History of the Human Sciences*, 1: 179–200.
____. (1998) 'Medicine, History and the Present', in C. Jones and R. Porter (eds) *Reassessing Foucault: Power, Medicine and the Body*, London and New York: Routledge.
—— (1996) 'Governing "advanced" liberal democracies', in A. Barry, T. Osborne and N. Rose (eds) *Foucault and Political Reason. Liberalism, Neo-liberalism and Rationalities of Government*, Chicago: The University of Chicago Press.
____. (1999) *Powers of Freedom. Reframing Political Thought*, Cambridge: Cambridge University Press.
____. (2007) *The Politics of Life Itself: Biomedicine, Power, and Subjectivity in the Twenty-First Century*, Princeton: Princeton University Press.
——. (2013) 'The human sciences in a biological age', *Theory, Culture & Society*, 30: 3–34.

____. and Novas, C. (2005) 'Biological Citizenship', in A. Ong and S. J. Collier (eds) *Global Assemblages: Technology, Politics, and Ethics as Anthropological Problems*, Malden: Blackwell.

____. and Abi-Rached, J. M. (2013) *Neuro: The New Brain Sciences and the Management of the Mind*, Princeton: Princeton University Press.

Rosenberg, C. E. (2006) 'Contested boundaries: psychiatry, disease, and diagnosis', *Perspectives in Biology and Medicine*, 49: 407–24.

Rosenthal, L. D. (2002) 'Requiring individuals to use mitigating measures in reasonable accommodation cases after the Sutton trilogy: putting the brakes on a potential runaway train', *South Carolina Law Review*, 54: 421.

Roskies, A. L. (2008) 'Neuroimaging and inferential distance', *Neuroethics*, 1: 19–30.

Rothblum, E. and Solovay, S. (eds) (2009) *The Fat Studies Reader*, New York and London: New York University Press.

Rousseau, P. (1999) 'Ascetics as Mediators and as Teachers', in J. Howard-Johnston and P. A. Hayward (eds) *The Cult of Saints in Late Antiquity and the Middle Ages: Essays on the Contribution of Peter Brown*, Oxford: Oxford University Press.

Roy, S. C. (2008) '"Taking charge of your health": discourses of responsibility in English-Canadian women's magazines', *Sociology of Health & Illness*, 30: 463–77.

Rubenson, S. (2002) 'Christian Asceticism and the Emergence of the Monastic Tradition', in V. L. Wimbush and R. Valantasis (eds) *Asceticism*, New York: Oxford University Press.

Rubin, C. T. (2008) 'What is the Good of Transhumanism?', in B. Gordijn and R. Chadwick (eds) *Medical Enhancement and Posthumanity*, New York: Springer-Verlag.

Russon, J. (2001) *The Self and Its Body in Hegel's 'Phenomenology of Spirit'*, Toronto, Buffalo and London: University of Toronto Press.

Rysst, M. (2010) '"Healthism" and looking good: body ideals and body practices in Norway', *Scandinavian Journal of Public Health*, 38: 71–80.

Sacks, O. (1990) *Seeing Voices: A Journey into the World of the Deaf*, New York: Harper Perennial.

Salecl, R. (2010) *Choice* (Big Ideas), London and New York: Profile Books.

Sant'Anna, D. (2001) *Corpos de passagem: ensaios sobre a subjetividade contemporânea*, São Paulo: Estação Liberdade.

Santosuosso, A. (2001) *Corpo e libertà. Una storia tra diritto e scienza*, Milano: Raffaelo Cortina Editore.

Sarasin, P. and Tanner, J. (eds) (1998) *Physiologie und Industrielle Gesellschaften: Studien zur Verwissenschaftlichung des Körpers im 19. und 20. Jahrhundert*, Frankfurt am Main: Suhrkamp.

Satava, R. M. (1995) 'Computers in surgery: telesurgery, virtual reality, and the new world order of medicine', *Contemporary Surgery*, 47: 204–208.

Savulescu, J. (2002) 'Deaf lesbians, "designer disability", and the future of medicine', *British Medical Journal*, 325: 771–73.

____. (2009) 'Refusing cochlear implants: is it child neglect?', available at http://blog.practicalethics.ox.ac.uk/2009/07/refusing-cochlear-implants-is-it-child-neglect/ (accessed 15 July 2009).

Sawday, J. (1996) *The Body Emblazoned: Dissection and the Human Body in Renaissance Culture*, London and New York: Routledge.

Saxton, M. (2010) 'Disability Rights and Selective Abortion', in L. J. Davis (ed) *The Disability Studies Reader*, 3rd edn, New York: Routledge.
Scarry, E. (1987) *The Body in Pain: The Making and Unmaking of the World*, Oxford: Oxford University Press.
Schaffer, R, Kuczynski, K. and Skinner, D. (2008) 'Producing genetic knowledge and citizenship through the internet: mothers, pediatric genetics, and cybermedicine', *Sociology of Health & Illness*, 30: 145–59.
Schatzlein, J. (2009) 'Christopher Reeve: 1952–2004', in R. M. Baird, S. E. Rosenbaum and S. K. Toombs (eds) *Disability: The Social, Political, and Ethical Debate*, Amherst: Prometheus Books.
Scheer, E. (2002) 'What Does an Avatar Want? Stelarc's E-motions', in J. Zylinska (ed) *The Cyborg Experiments: The Extensions of the Body in the Media Age*, London and New York: Continuum.
Scheper-Hughes, N. (2005) 'The Last Commodity: Post-human Ethics and the Global Traffic in "Fresh" Organs', in A. Ong and S. J. Collier (eds) *Global Assemblages: Technology, Politics, and Ethics as Anthropological Problems*, Malden: Blackwell.
Schessler-Jandreau, I. (2008) 'Fat America: A Historical Consideration of Diet and Weight Loss in the US', in E. E. Peterson (ed) *Communication and Public Policy: Proceedings of the 2008 International Colloquium on Communication*, available at http://scholar.lib.vt.edu/ejournals/ICC/2008/ICC2008Schessler.pdf (accessed 16 June 2009).
Schluchter, W. (ed) (1988) *Max Webers Sicht des Okzidentalen Christentums*, Frankfurt: Suhrkamp.
Schrader, U. (2007) 'The moral responsibility of consumers as citizens', *International Journal of Innovation and Sustainable Development*, 2: 79–96.
Schürmann, V. (2003) 'Die Bedeutung der Körper. Literatur zur Körper-Debatte – eine Auswahl in systematischer Absicht', *Allgemeine Zeitschrift für Philosophie*, 28(1): 51–69.
Schweik, S. M. (2009) *The Ugly Laws. Disability in Public*, New York and London: New York University Press.
Scott, W. R. et al (2000) *Institutional Change and Healthcare Organizations: From Professional Dominance to Managed Care*, Chicago: University of Chicago Press.
Seidel, A. (2009) *Inhuman Thoughts: Philosophical Explorations of Posthumanity*, Lanham: Lexington Books.
Segal, A. (1979) 'Les progrès de l'exploration visuelle des organes aux XVIIe et XVIIIe siècles', *Histoire des sciences médicales*, 13(1): 395–406.
Sfez, L. (1995) *La santé parfaite. Critique d'une nouvelle utopie*: Paris: Seuil.
Shakespeare, R. (2006) *Disability Rights and Wrongs*, Abingdon and New York: Routledge.
Sharp, L. A. (2000) 'The commodification of the body and its parts', *Annual Review of Anthropology*, 29: 287–328.
_____. (2006) *Strange Harvest: Organ Transplants, Denatured Bodies and the Transformed Self*, Berkeley: California University Press.
Shaw, S. (2002) 'Why courts cannot deny ADA protection to plaintiffs who do not use available mitigating measures for their impairments', *California Law Review*, 90: 1981–2046.

Shildrick, M. (2010) 'Some reflections on the socio-cultural and bioscientific limits of bodily integrity', *Body & Society*, 16: 11–22.

____. and Price, J. (1999) 'Openings on the Body: A Critical Introduction', in J. Price and M. Shildrick (ed) *Feminist Theory and the Body: A Reader*, London: Routledge.

Shilling, C. (2012) *The Body and Social Theory*, 3rd edn, London: Sage Publications.

Shusterman, R. (2000) *Performing Live: Aesthetic Alternatives for the Ends of Art*, Ithaca and London: Cornell University Press.

_____. (2003) 'Somaesthetics and *The Second Sex*: a pragmatist reading of a feminist classic', *Hypatia*, 18(4): 106–36.

_____. (2012) *Thinking Through the Body. Essays in Somaesthetics*, Cambridge: Cambridge University Press.

Siebers, T. (2001) 'Disability in theory: from social constructionism to the new realism of the body', *American Literary History*, 13: 737–54.

____. (2010) 'In the Name of Pain', in J. M. Metzl and A. Kirkland (eds) *Against Health: How Health Became the New Morality*, New York: New York University Press.

Silverman, C. (2008) 'Fieldwork on another planet: social science perspectives on the autism spectrum', *BioSocieties*, 3: 325–41.

Silvers, A., Wasserman, D. and Mahowald, M. B. (1998) *Disability, Difference, Discrimination. Perspective on Justice in Bioethics and Public Policy*, Lanham: Rowman and Littlefield.

Simpson, J. R. (ed) (2012) *Neuroimaging in Forensic Psychiatry: From the Clinic to the Courtroom*, Oxford, Chichester and Hoboken: Wiley-Blackwell.

Sinclair, J. (2005) 'Autism network international: the development of a community and its culture' available at http://web.syr.edu/$jisincla/History_of_ANI.html (accessed 13 December 2007).

Singer, J. (1999) 'Why Can't You be Normal for once in Your Life? From a "Problem with No Name" to the Emergence of a New Category of Difference', in M. Corker and S. French (eds) *Disability Discourse*, Buckingham, UK: Open University Press.

——. (2007) 'Light and dark: correcting the balance', available at http://www.neurodiversity.com.au/ (accessed 10 January 2010).

Singh, I. (2012a) 'Human development, nature and nurture: working beyond the divide', *BioSocieties*, 7: 308–21.

Singh, I. (2012b) 'Brain talk: power and negotiation in children's discourse about self, brain and behaviour', *Sociology of Health & Illness*, 35: 813–827.

Skrabanek, P. (1994) *The Death of Human Medicine and the Rise of Coercive Healthism*, London: Social Affairs Unit.

Slaby, J. and Choudhury, S. (2012) 'Proposal for a Critical Neuroscience', in S. Choudhury and J. Slaby (eds) *Critical Neuroscience: A Handbook of the Social and Cultural Contexts of Neuroscience*, Oxford: Wiley-Blackwell.

Smart, A. (2003) 'Reporting the dawn of the Post-Genomic Era: who wants to live forever?', *Sociology of Health and Illness*, 25: 24–49.

Smith, M. and Morra, J. (eds) (2006) *The Prosthetic Impulse: From a Posthuman Present to a Biocultural Future*, Cambridge, Mass: The MIT Press.

Snell, B. (1953) *The Discovery of the Mind. The Greek Origins of European Thought*, trans. T. G. Rosenmeyer, Oxford: Blackwell.

Sokolow, J. A. (1983) *Eros and Modernization. Sylvester Graham, Health Reform, and the Origins of Victorian Sexuality in America*, London and Toronto: Associated Universities Press.

Sparrow, R. (2005) 'Defending deaf culture: the case of cochlear implants', *The Journal of Political Philosophy*, 13: 135–52.

Spriggs, M. (2002) 'Lesbian couple create a child who is deaf like them', *Journal of Medical Ethics*, 28: 283.

Stafford, B. M. (1991) *Body Criticism: Imaging the Unseen in Enlightenment Art and Medicine*, Cambridge, Mass: MIT Press.

____. (1996) *Good Looking: Essays on the Virtue of Images*, Cambridge, Mass: MIT Press.

Stark, C. A. (2010) 'Respecting Human Dignity: Contract Versus Capabilities', in E. F. Kittay and L. Carlson (eds) *Cognitive Disability and its Challenge to Moral Philosophy*, Malden and Oxford: Wiley-Blackwell.

Starobinski, J. (1985) *Montaigne in Motion*, trans. A. Goldhammer, Chicago and London: University of Chicago Press.

——. (2003) *Action and Reaction. The Life and Adventures of a Couple*, trans. Sophie Hawkes, New York: Zone Books.

Stelarc (1991) 'Prosthetics, robotics and remote existence. Post-evolutionary strategies', *Leonardo*, 24: 591–94.

____. (1998) 'From Psycho-body to Cyber-systems: Images as Post-human Entities', in J. Dixon and E. Cassidy (eds) *Virtual Futures: Cybererotics, Technology and Post-human Pragmatism*, London: Routledge.

____. (2000) 'Parasite Visions: Alternate, Intimate and Involuntary Experiences', in M. Featherstone (ed) *Body Modification*, London: Sage Publications.

____. (2002) 'Towards a Compliant Coupling: Pneumatic Projects, 1998–2001', in J. Zylinska (ed) *The Cyborg Experiments: The Extensions of the Body in the Media Age*, London and New York: Continuum.

Stern, M. (2006) 'Dystopian anxieties versus utopian ideals: medicine from Frankenstein to the Visible Human Project and Body Worlds', *Science as Culture*, 15: 61–84.

Stoff, H. (1999) 'Diskurse und Erfahrungen. Ein Rückblick auf die Körpergeschichte der neunziger Jahre', *Zeitschrift für Sozialgeschichte des 20. und 21. Jahrhunderts*, 14: 142–60.

Strauss, E. (1978) [1935] *Vom Sinn der Sinne*, Berlin: Springer.

Sullivan, N. (2002) 'Fleshly (dis)figuration, or how to make the body matter', *International Journal of Critical Psychology*, 5: 12–29.

Sunder Rajan, K. (2006) *Biocapital: The Constitution of Postgenomic Life*, Durham: Duke University Press.

Sunkaria, R. K., Kumar, V. and Chandra Saxena, S. (2010) 'A comparative study on spectral parameters of HRV in yogic and non-yogic practitioners', *Int. J. Medical Engineering and Informatics*, 2: 1–14.

Sweetman, P. (2000) 'Anchoring the (Postmodern) Self? Body Modification, Fashion and Identity', in M. Featherstone (ed) *Body Modification*, London: Sage Publications.

Szondi, P. (1986) *On Textual Understanding and Other Essays*, Minneapolis: University of Minnesota Press.

Tambornino, J. (2002) *The Corporeal Turn: Passion, Necessity, Politics*, Lanham: Rowman & Littlefield Publishers, Inc.

Tannen, S. (undated) 'Mental fitness – exercises for the brain', available at http://bellydoc.com/articles/article7.htm (accessed 25 March 2008).

Taylor, C. (1989) *Sources of the Self: The Making of Modern Identity*, Cambridge, Mass: Harvard University Press.

Taylor, D. and Vintges, K. (eds) (2004) *Feminism and the Final Foucault*, Urbana and Chicago: University of Illinois Press.
Thacker, E. (2000) 'Performing the Technoscientific Body: RealVideo Surgery and the Anatomy Theater', in M. Featherstone (ed) *Body Modification*, London: Sage Publications.
Tierney, S. (2001) 'A reluctance to be defined "disabled". How can the social model of disability enhance understanding of anorexia?', *Disability & Society*, 16: 749–64.
Tierney, T. F. (1998) 'Anatomy and Governmentality: A Foucauldian Perspective on Death and Medicine in Modernity', *Theory & Event*, 2(1): http://calliope.jhu.edu:80/journals/theory_&_event
Todes, S. (2001) *Body and World*, Cambridge, Mass: MIT Press.
Toffoletti, K. (2007) *Cyborgs and Barbie Dolls. Feminism, Popular Culture and the Posthuman Body*, New York: Tauris & Co.
Toombs, S. K. (ed) (2001) *Handbook of Phenomenology and Medicine*, Dordrecht, Boston and London: Kluwer Academic Publishers.
Treichler, P., Cartwright, L. and Penley, C. (eds) (1998) *The Visible Woman: Imaging Technologies, Gender and Science*, New York: New York University Press.
Tremain, S. (2005) 'Foucault, Governmentality, and Critical Disability Theory. An Introduction', in S. Tremain (ed) *Foucault and the Government of Disability*, Ann Arbor: The University of Michigan Press.
Tronzo, W. (ed) (2009) *The Fragment. An Incomplete History*, Los Angeles: Getty Publications.
Tucker, B. P. (1997) 'The ADA and deaf culture: contrasting precepts, conflicting results', *Annals of the American Academy of Political and Social Science*, 549: 24–36.
____. (1998) 'Deaf culture, cochlear implants, and elective disability', *Hastings Center Report*, 28: 6–14.
Turner, B. (2008 [1984]) *Body and Society: Explorations in Social Theory*, 3rd edition, Los Angeles and London: Sage.
____. (2000) 'The Possibility of Primitiveness: Towards a Sociology of Body Marks in Cool Societies', in M. Featherstone (ed) *Body Modification*, London: Sage Publications.
____. (2002) 'The Anatomy Lesson: A Note on the Merton Thesis', in Bryan S. Turner, *Regulating Bodies. Essays in Medical Sociology*, London and New York: Routledge.
____. (2006) *Vulnerability and Human Rights*, University Park: Pennsylvania State University Press.
____. (2012a) 'Introduction: The Turn of The Body', in B.S. Turner (ed) *Routledge Handbook of Body Studies*, London: Routledge.
____. (2012b) 'Embodied Practice', in B. S. Turner (ed) *Routledge Handbook of Body Studies*, London: Routledge.
Turner, T. (2001) 'Bodies and anti-bodies: flesh and fetish in contemporary social theory', in T. J. Csordas (ed) *Embodiment and Experience. The Existential Ground of Culture and Self*, Cambridge: Cambridge University Press.
Valantasis, R. (2008) *The Making of the Self. Ancient and Modern Asceticism*, Eugene, Oregon: Cascade Books.
Valverde, M. (1998) *Diseases of the Will: Alcohol and the Dilemmas of Freedom*, Cambridge: Cambridge University Press.
____. (2004) 'Experience and Truth Telling in a Post-humanist World: A Foucauldian Contribution to Feminist Ethical Reflection', in D. Taylor and K. Vintges

(eds) *Feminism and the Final Foucault*, Urbana and Chicago: University of Illinois Press.
van Dijck, J. (2000) 'Digital cadavers: the visible human project as anatomical theater', *Studies in History and Philosophy of Science*, 31: 271–85.
____. (2001) 'Bodyworlds: the art of plastinated cadavers', *Configurations*, 9: 99–126.
____. (2005) *The Transparent Body. A Cultural Analysis of Medical Imaging*, Seattle: University of Washington Press.
van Vall, R. and Zwijnenberg, R. (eds) (2009). *The Body Within: Art, Medicine and Visualization*, Leiden and Boston: Brill.
van Wyhe, J. (2004) *Phrenology and the Origins of Victorian Scientific Naturalism*, Aldershot: Ashgate.
Vazquez Garcia, F. (2005) *Tras la autoestima. Variaciones sobre el yo en la modernidad tardía*, San Sebastian: Gakoa.
Vesalius, A. (1998) [1543] *On the Fabric of the Human Body, Book I, The Bones and Cartilages*. A translation of De Humani Corporis Fabrica Libri Septem, trans. W. F. Richardson, in collaboration with J. B. Carman, San Francisco: Norman Publishing.
Vidal, F. (2009) 'Brainhood, anthropological figure of modernity', *History of the Human Sciences*, 22: 5–36.
____. (2011) 'La neuroesthétique, un esthétisme scientiste', *Revue d'histoire des sciences humaines*, 25: 239–64.
Voelke, A. J. (1973) *L'idée de volonté dans le stoïcisme*, Paris: Presses Universitaires de France.
Vrecko, S. (2006) 'Folk neurology and the remaking of identity', *Molecular Interventions*, 6: 300–303.
——. (2011) 'On the Political Economy of the Gambling Brain', in F. Ortega and F. Vidal (eds) *Neurocultures: Glimpses into an Expanding Universe*, Frankfurt and New York: Peter Lang.
Von Hagens, G. (2008) *BodyWorlds: The Original Exhibition of Real Human Bodies*, 13th edition, Heidelberg: Arts & Sciences Verlagsgesellschaft.
Waldby, C. (1997) 'Revenants: The Visible Human Project and the Digital Uncanny', *Body Society*, 3: 1–16.
——. (2000) *The Visible Human Project. Informatic Bodies and Posthuman Medicine*, London and New York: Routledge.
Waldby, C. and Mitchell, R. (2006) *Tissue Economies: Blood, Organs, and Cell Lines in Late Capitalism*, Durham: Duke University Press.
Waldby, C., Rosengarten, M., Treloar, C. and Fraser, S. (2004) 'Blood and bioidentity: ideas about self, boundaries and risk among blood donors and people living with hepatitis C', *Social Science and Medicine*, 59: 1461–71.
Waldenfels, B. (2000) *Das Leibliche Selbst. Vorlesungen zur Phänomenologie des Leibes*, Frankfurt am Main: Suhrkamp.
Wann, M. (2009) 'Foreword: Fat Studies: An Invitation to Revolution', in E. Rothblum and S. Solovay (eds) *The Fat Studies Reader*, New York and London: New York University Press.
Ware, K. (2002) 'The Way of the Ascetics: Negative or Affirmative?', in V. L. Wimbush and R. Valantasis (eds) *Asceticism*, New York: Oxford University Press.
Weber, M. (2003) [1905]. *The Protestant Ethic and the Spirit of Capitalism*, trans. Talcott Parsons, Mineola, NY: Dover.

Wehling, P. (2011) 'Biology, Citizenship and the Government of Biomedicine: Exploring the Concept of Biological Citizenship', in U. Bröckling, S. Krasmann and T. Lemke (eds) *Governmentality: Current Issues and Future Challenges*, New York: Routledge.

Weinberg, D. (2012) 'Social Constructionism and the Body', in B. S. Turner (ed) *Routledge Handbook of Body Studies*, London: Routledge.

Weiss, G. (1999a) 'The Abject Borders of the Body Image', in G. Weiss and H. F. Haber (eds) *Perspectives on Embodiment: The Intersections of Nature and Culture*, London and New York: Routledge.

____. (1999b) *Body Images: Embodiment as Intercorporeality*, New York and London: Routledge.

Weisz, G. (2006) *Divide and Conquer. A Comparative History of Medical Specialization*, Oxford and New York: Oxford University Press.

Welton, D. (1999) 'Soft, Smooth Hands: Husserl's Phenomenology of the Lived-body', in D. Welton (ed) *The Body: Classic and Contemporary Readings*, Bodmin, Cornwall: Blackwell Publishers.

Wendell, S. (1996) *The Rejected Body: Feminist Philosophical Reflections on Disability*, New York: Routledge.

Wenner, S. (2001) 'Ganzer oder zerstückelter Körper. Über die Reversibilität von Körperbildern', in C. Benthien and C. Wulf (eds) *Körperteile. Eine kulturelle Anatomie*, Hamburg: Rowohlt.

——. (2002) 'Unversehrter Leib im "Reich der Zwecke": Zur Genealogie des *Cyborgs*', in A. Barkhaus and A. Fleig (eds) *Grenzverläufe. Der Körper als Schnitt-Stelle*, München: Wilhelm Fink Verlag.

Whyte, S. R. (2009) 'Health identities and subjectivities: the ethnographic challenge', *Med Anthropol Quarterly*, 23: 6–15.

Williams, S. J. (1997) 'Modern medicine and the "uncertain body": from corporeality to hyperreality?', *Social Science and Medicine*, 45: 1041–49.

____. (2003) *Medicine and the Body*, London: Sage Publications.

____. and Bandelow, G. (1998) *The Lived Body: Sociological Themes, Embodied Issues*, London and New York: Routledge.

Williams, J. (2009) 'Christopher Reeve's Super Bowl Ad Scored a Touchdown: But It has Provoked a Surprisingly Negative Reaction among Disabled Groups. Why?', in R. M. Baird, S. E. Rosenbaum and S. K. Toombs (eds) *Disability: The Social, Political, and Ethical Debate*, Amherst: Prometheus Books.

Wilson, E. A. (2008) 'Organic Empathy: Feminism, Psychopharmaceuticals and the Embodiment of Depression', in S. Alaimo and S. Hekman (eds) *Material Feminisms*, Bloomington: Indiana University Press.

Wimbush, V. L. (1990) 'Introduction', in V. L. Wimbush (ed) *Ascetic Behavior in Greco-Roman Antiquity: A Sourcebook*, Minneapolis: Fortress Press.

____. and Valantasis, R. (2002) 'Introduction', in V. L. Wimbush and R. Valantasis (eds) *Asceticism*, New York: Oxford University Press.

Winnicott, D.W. (1990) *The Maturational Processes and the Facilitating Environment: Studies in the Theory of the Emotional Development*, London: Karnac Books.

____. (1992) *Collected Papers: Through Paediatrics to Psycho-Analysis*, London: Karnac Books.

____. (1999), *Playing and Reality*, New York: Routledge.

Winter, A. and Winter, R. (1987) *Build Your Brain Power: The Latest Techniques to Preserve, Restore, and Improve Your Brain's Potential*, New York: St Martin's Press.

Witz, A. (2000) 'Whose body matters? Feminist sociology and the corporeal turn in sociology and feminism', *Body & Society*, 6: 21–24.

Wohlrab, S., Stahl, J. and Kappeler, P. M. (2007) 'Modifying the body: motivations for getting tattooed and pierced', *Body Image*, 4: 87–95.

Wolbarst, A. B. (1999) *Looking Within. How X-Ray, CT, MRI, Ultrasound and Other Medical Images Are Created and How They Help Physicians Save Lives*, Berkeley and Los Angeles: University of California Press.

Wolf, J. B. (2010) 'Against Breastfeeding (Sometimes)', in J. M. Metzl and A. Kirkland (eds) *Against Health. How Health Became the New Morality*, New York and London: New York University Press.

____. (2011) *Is Breast Best? Taking on the Breastfeeding Experts and the New High Stakes of Motherhood*, New York and London: New York University Press.

Wolfe, C. (ed) (2010) *What is Posthumanism?* Posthumanities series, Minneapolis: University of Minnesota Press.

Wong, S. I. (2010) 'Duties of Justice to Citizens with Cognitive Disabilities', in E. F. Kittay and L. Carlson (eds) *Cognitive Disability and its Challenge to Moral Philosophy*, Malden and Oxford: Wiley-Blackwell.

Yates, C. (2010) 'Stations of the self: aesthetics and ascetics in Foucault's conversion narrative', *Foucault Studies*, 8: 78–99.

Yehya, N. (2001) *El cuerpo transformado*, México D.F.: Paidós.

Young, C. (2002) 'Sound judgment: does curing deafness really mean genocide?' *Reason*, 33: 20–21.

Young, I. M. (2005) *On Female Body Experience: 'Throwing Like a Girl' and Other Essays* Oxford: Oxford University Press.

____. (2007) 'Structural Injustice and the Politics of Difference', in K. A. Appiah, S. Benhabib, I. M. Young and N. Fraser (eds) *Justice, Governance, Cosmopolitanism, and the Politics of Difference: Reconfigurations in a Transnational World.* Distinguished W. E. B. Du Bois Lectures 2004/2005, Berlin: Der Präsident der Humboldt-Universität zu Berlin.

Young, K. (1997) *Presence in the Flesh. The Body in Medicine*, Cambridge, Mass: Harvard University Press.

Yourcenar, M. (2000) *Memoirs of Hadrian*, trans. G. Frick, New York: Farrar, Straus and Giroux.

Zaner, R. M. (1992) 'Parted Bodies, Departed Souls: The Body in Ancient Medicine and Anatomy', in D. Leder (ed) *The Body in Medical Thought and Practice*, Dordrecht, Boston and London: Kluwer Academic Publishers.

Žižek, S. (2002) *Welcome to the Desert of the Real!: Five Essays on September 11 and Related Dates*, London: Verso.

Zurbrugg, N. (2000) 'Marinetti, Chopin, Stelarc and the Auratic Intensities of the Postmodern Techno-body', in M. Featherstone (ed) *Body Modification*, London: Sage Publications.

Zylinska, J. (2002) '"The Future . . . is Monstrous": Prosthetics as Ethics', in J. Zylinska (ed) *The Cyborg Experiments: The Extensions of the Body in the Media Age*, London and New York: Continuum.

____. and Hall, G. (2002) 'Probings: An Interview with Stelarc', in J. Zylinska (ed). *The Cyborg Experiments: The Extensions of the Body in the Media Age*, London and New York: Continuum.

****

Auton (Guardian ad litem of) v British Columbia (Attorney General) 2004 SCC 78.
Bertinetti v Joy Mining Machinery, 231 F. Supp. 2d 828 (S.D. Il. 2002).
Coghlan v H.J. Heinz Co., 851 F.Supp. 808 (N.D. Tex. 1994).
Edwards v Blomeley [2002] NSWSC 460.
Eldridge v British Columbia (Attorney General) [1997] 3 SCR 624.
Gleitman v Cosgrove, 49 N.J. 22, 227 A.2d 689 (1967).
Granovski v Canada (Minister of Employment and Immigration) [2000] 1 SCR703.
Harriton v Stephens [2002] NSWSC 461.
Nawrot v CPC Intern., 277 F.3d 896, 904 (7th Cir. 2002).
Pangalos v Prudential Insurance Company of America (1997) United States Court of Appeals, Third Circuit 118 F.3d 1577.
Pennhurst State School and Hospital v Halderman, 451 U.S. 1 (1981) 457 U.S. 1131 (1984).
Sutton v United Airlines, Inc., 527 U.S. 471 (1999).
Thornton v McClatchy, 261 F. 3d 789, 798 (9th Cir. 2001).
Waller v James [2002] NSWSC 462.

Index

For Product Safety Concerns and Information please contact our EU representative GPSR@taylorandfrancis.com
Taylor & Francis Verlag GmbH, Kaufingerstraße 24, 80331 München, Germany

www.ingramcontent.com/pod-product-compliance
Ingram Content Group UK Ltd.
Pitfield, Milton Keynes, MK11 3LW, UK
UKHW022025061126
477531UK00002B/47
9780415593229